Human Anatomy

in Full Color

Illustrations by John Green

Text by Dr. John W. Harcup
S.B.St.J., M.R.C.G.P., M.R.C.S., D.Obst. R.C.O.G.
Fellow of the Royal Society of Medicine

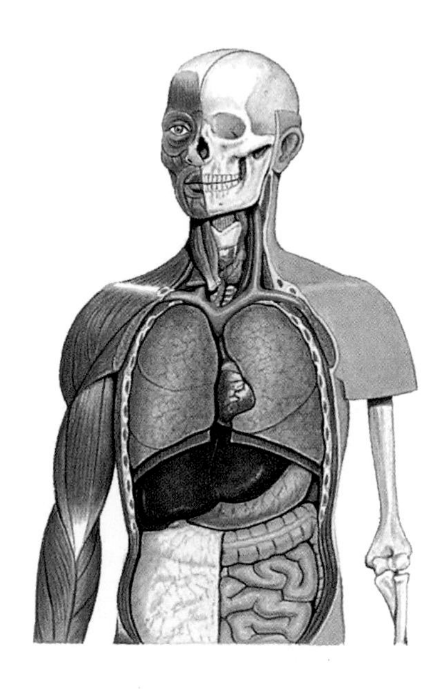

DOVER PUBLICATIONS, INC.
MINEOLA, NEW YORK

Bibliographical Note

Human Anatomy in Full Color is a new work, first published by Dover Publications, Inc., in 1996.

Library of Congress Cataloging-in-Publication Data

Green, John, 1948–
Human anatomy in full color / illustrations by John Green ; text by John W. Harcup.
p. cm.
Includes index.
ISBN 0-486-29065-4 (pbk.)
1. Human anatomy—Atlases. I. Harcup, John W. II. Title.
QM25.G74 1996
611'.0022'2—dc20 96-42325
CIP

Manufactured in the United States by LSC Communications
29065412 2016
www.doverpublications.com

INTRODUCTION

The human body is an amazing structure, so complex that, even today, not all its functions are fully understood. Certain cells are directed by genes to divide into organs with specialized functions. The senses—sight, hearing, taste, touch and smell—connect us to the outside world. The body is capable of detecting minute changes in the environment. It can adapt to different conditions and extremes of heat and cold. Quickly, it can prepare for flight or fight to protect itself from danger. However, when the enemy is within, as tiny as a bacterium or virus, the human body has its own protective mechanism in the form of white blood cells and antibodies that kill the invader. Many ingested poisons are made harmless in the liver; waste products are efficiently excreted by the kidneys and alimentary canal.

The brain can appreciate beauty, devise art forms, compose music and create pictures, solve puzzles, design clothes, invent rockets to probe space and program computers. With specialized appendages—the hands—the body can make models and sculptures, paint paintings, knit, embroider exquisite garments, communicate by writing and drive cars and planes with incredible accuracy. The legs provide a means of locomotion. The ability to walk upright allows the arms freedom. John Green has drawn the human body to make it clearly understandable. In the text, I have tried to explain the body and to answer some of the questions I have been asked, as a doctor, on this subject.

John W. Harcup
Malvern, England
1996

CONTENTS

The Skeleton

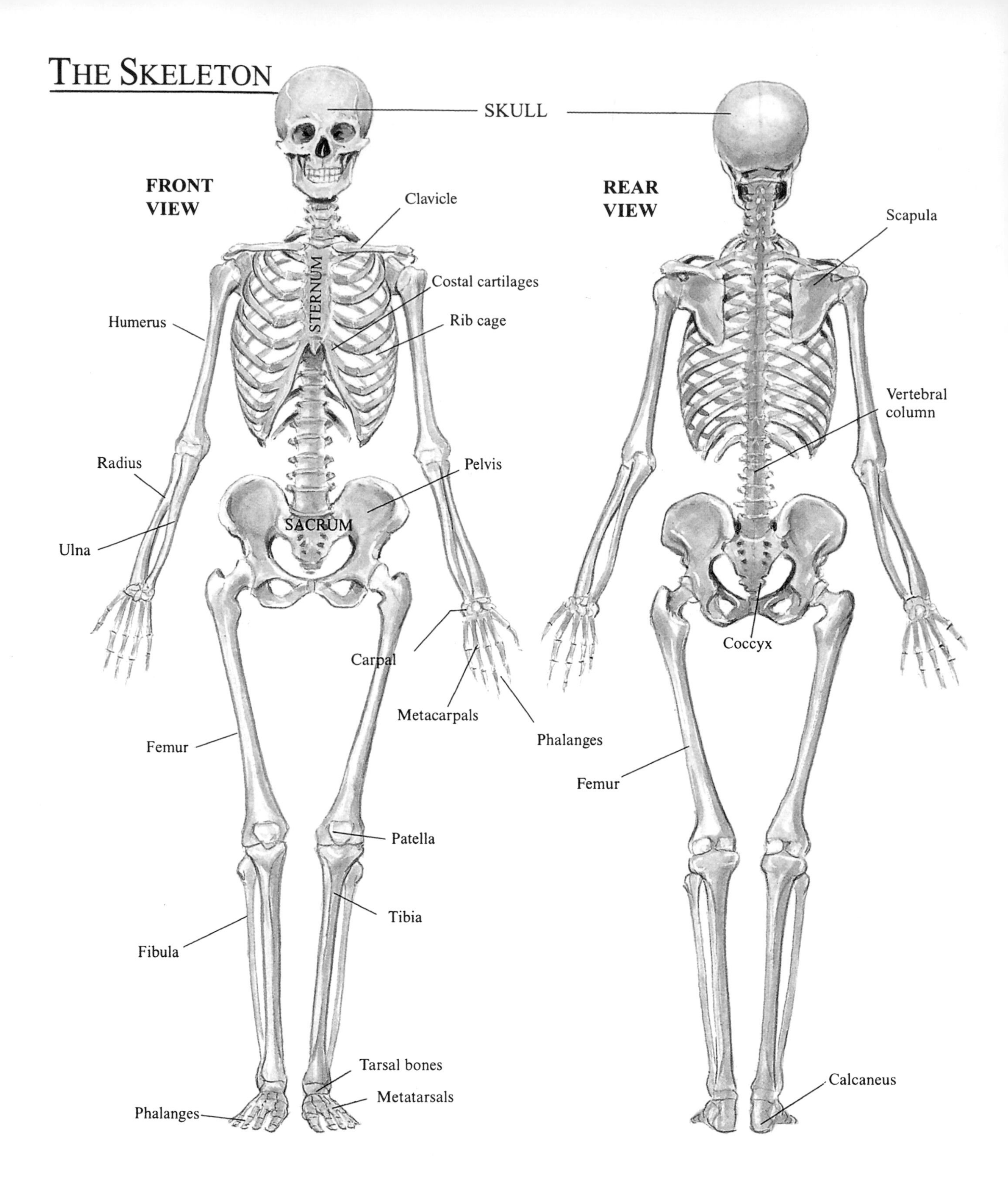

The 206 bones in the skeleton form the framework of the body, accounting for one seventh of an adult male's weight. The 80 bones of the head, face, neck and trunk are called the axial skeleton. The appendicular skeleton, comprising the limbs or appendages, is made up of 32 bones in each adult upper limb and 31 in each lower limb. Sesamoid bones are connected to a tendon rather than to another bone; everybody has at least two—the patella (kneecap) of each leg—but some people have them in other sites such as the foot or over the first joint in the thumb. (If present, these are extra bones and are always small.) The largest bone in the body is the femur, the smallest are the ossicles in the middle ear.

Bones have several important functions. They are the support system of the body for soft tissues and anchorage points for muscles. To allow movement of the body, individual bones must be able to move in relationship to one another. This happens at joints, where the bones articulate with each other, the surfaces that touch being termed articular surfaces.

Groups of bones form protective cages: The skull protects the

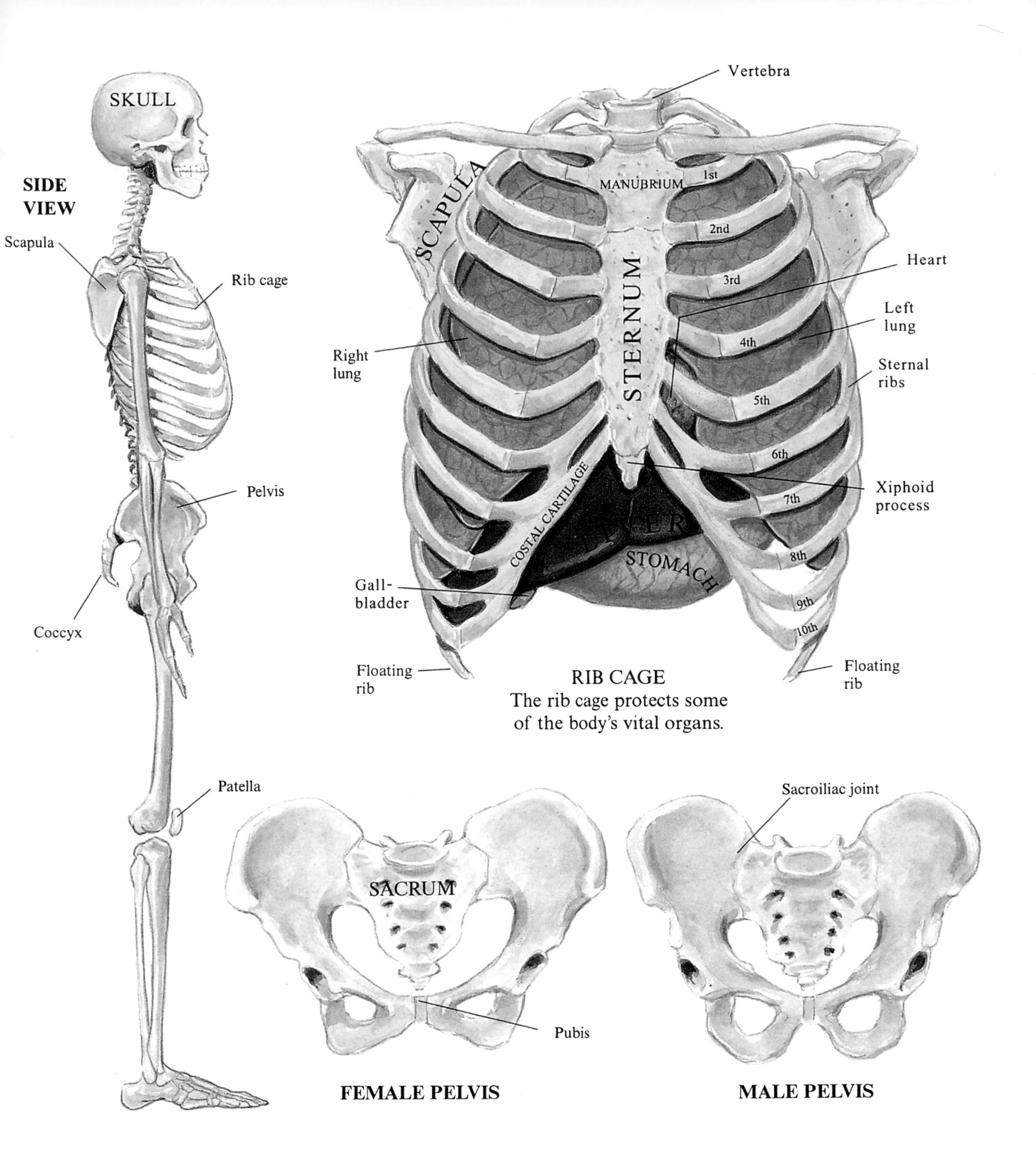

RIB CAGE
The rib cage protects some of the body's vital organs.

brain, and the lungs and heart are enclosed by the bones of the vertebral column (spine), ribs and sternum (breastbone). There are 12 pairs of ribs; the lower two pairs of ribs "float" and are not connected to the "gristle" or cartilage that composes the junction between the ribs and breastbone. In the pelvis the bone arrangement protects vital organs of the lower abdomen, but the female pelvis is wider than the male to allow space for a baby to pass through when it is born. The sacrum ends in the coccyx, which is a residual tail.

Inside bones, red and white blood cells and platelets are manufactured and essential elements such as calcium and phosphate are stored.

Bones & Joints

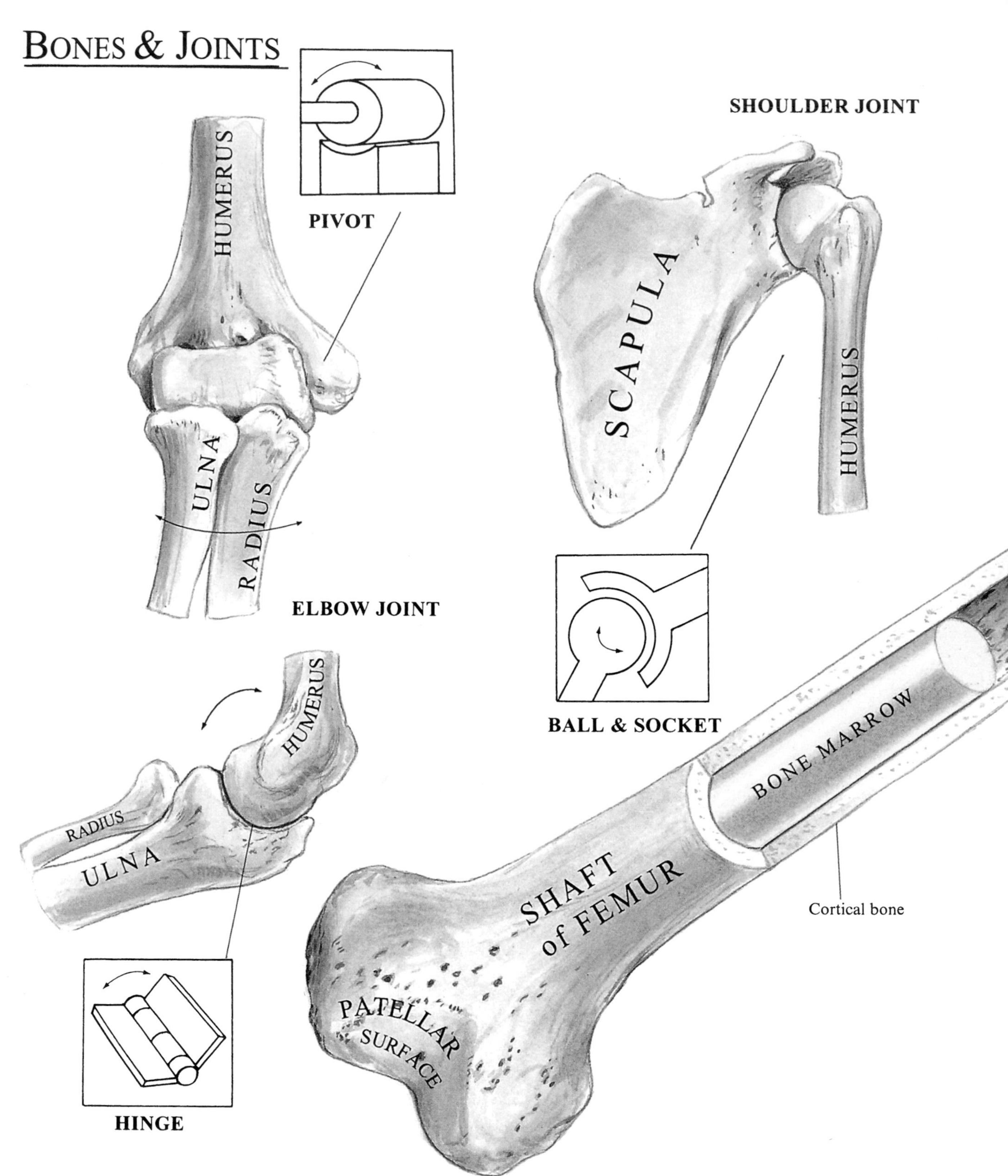

There are three kinds of bone—long (e.g., thigh and arm); flat (the shoulders, skull, ribs and pelvis) and irregular (the spinal column, ankle and wrist, although the last are sometimes called short bones). Bones are covered by a tough sheath, the periosteum, which contains many small nerve endings. This is why a blow on a bone such as the shin is so painful. The two types of bone, spongy (cancellous) and cortical, are composed of Haversian canals carrying nerves, blood vessels and lymphatics, surrounded by plates of bone and spaces containing bone cells—osteoblasts and osteocytes, which produce bone, and osteoclasts, which reabsorb it. Even in adults, 15 percent of bone may be remodeled in a year. Spongy bone has the appearance of a sponge as the canals are larger and the spaces contain bone marrow, comprised of blood and fat cells. In long bones, red blood cells are manufactured in the red bone marrow, which predominates in childhood, but in adults, this is replaced mainly by yellow, fatty marrow in the medullary cavity.

Bones are formed from calcium and phosphorus derived from diet and combined to form calcium phosphate. Long bones grow as new bone is deposited by the layer beneath the periosteum and at bands of cartilage, called epiphyseal plates, near the articular surfaces, until full growth is reached between the ages of 18 and 25, when bone replaces this cartilage. The age at which a bone stops growing varies from one bone to another.

A joint is formed where two bones meet. The surfaces on which

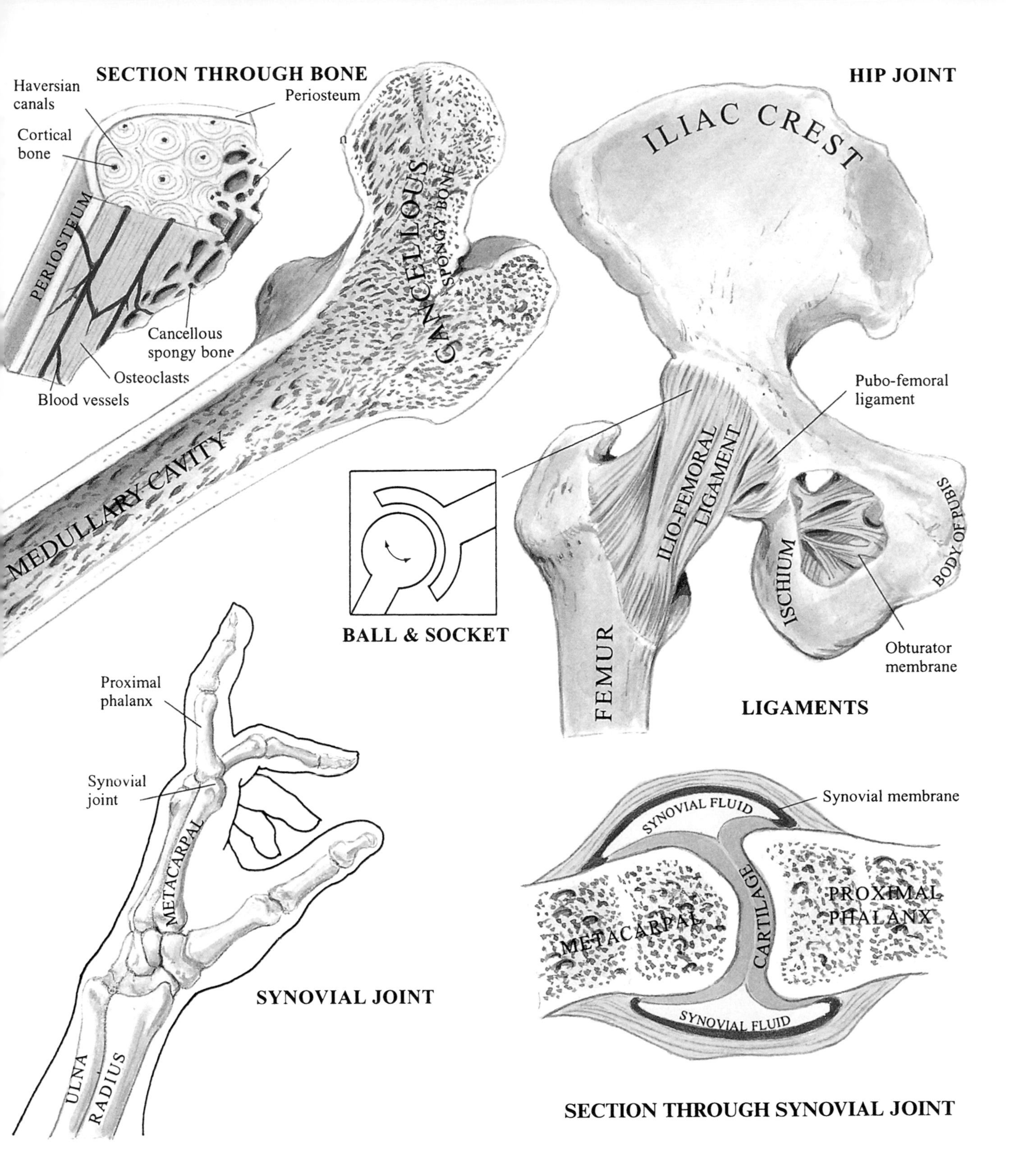

they move on one another are articular surfaces. They are covered in articular cartilage, which acts as a cushion. There are three types of joints. Fibrous joints are fixed and immovable, such as the bones of the skull. Cartilaginous joints are slightly movable joints in which a plate of fibrocartilage separates the bones, as in the breastbone (sternum) and the vertebral bodies of the spinal column. Synovial joints are freely movable and lined by a membrane producing a lubricant, synovial fluid. There are seven varieties of synovial joint. A hinge joint, like the elbow, moves up and down like the lid of a box. A pivot joint allows rotation only, as in the radius and ulna bones at the elbow, or the second bone in the neck. In a plane joint the surfaces are flat, as in the wrist and foot. Ball and socket joints, such as the shoulder and hip, can move in all directions. Condyloid joints are modified ball-and-socket joints that cannot move upward, such as the fingers and knuckle joints. An ellipsoid joint is another modified ball-and-socket joint in the wrist involving the carpal and radius bones, capable of all movements except rotation. A saddle joint, such as the thumb, moves on two axes at right angles to each other. Joints are surrounded by a capsule, which is further strengthened by ligaments.

THE SKULL

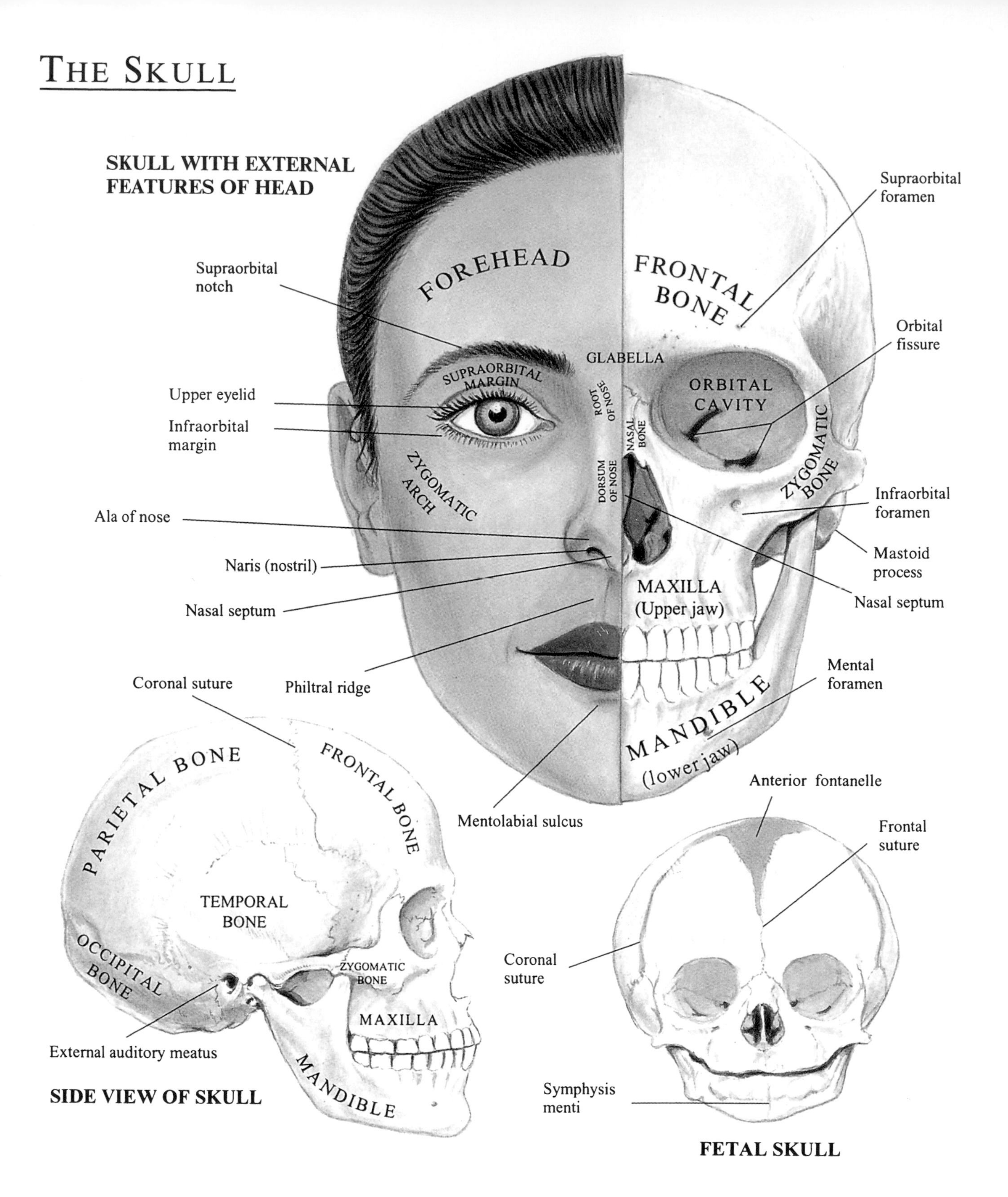

The facial appearance of any person depends partly on the underlying size and shape of the skull, which is the skeleton of the head and face, and the muscles attached to it. Only the mandible of the jaw moves; the other 21 bones have fused together by adulthood. Eight of these bones are paired, five are unpaired and the areas of fusion are known as sutures. The infant skull differs from the adult: It is larger in comparison with the rest of the body. Infant sutures are very wide at birth and allow "molding" or shaping of the skull for a damage-free passage through the birth canal. During the first 18 months of life the gaps can be felt in certain places, especially the diamond-shaped anterior fontanelle at the junction of the frontal and parietal bones. Another space—between the occipital and parietal bones—can be felt during a baby's first two months. Sometimes, in disease, the bones fuse together too soon or stay apart too long.

The only movable joint in the head is the temporomandibular, where the mandible (lower jaw) forms a hinge with the temporal bone. It can move in three planes, placing the jaw up, down, backward and forward as well as side to side. There are three specialized openings in the skull for ears, eyes and nose. The eye nestles in the orbital cavity, which is protected by a bony rim. Blows to the eye may fracture the zygomatic bone and maxilla (part of the upper jaw) rather than injure the eye itself. The nose has a root and roof of bone but the lower

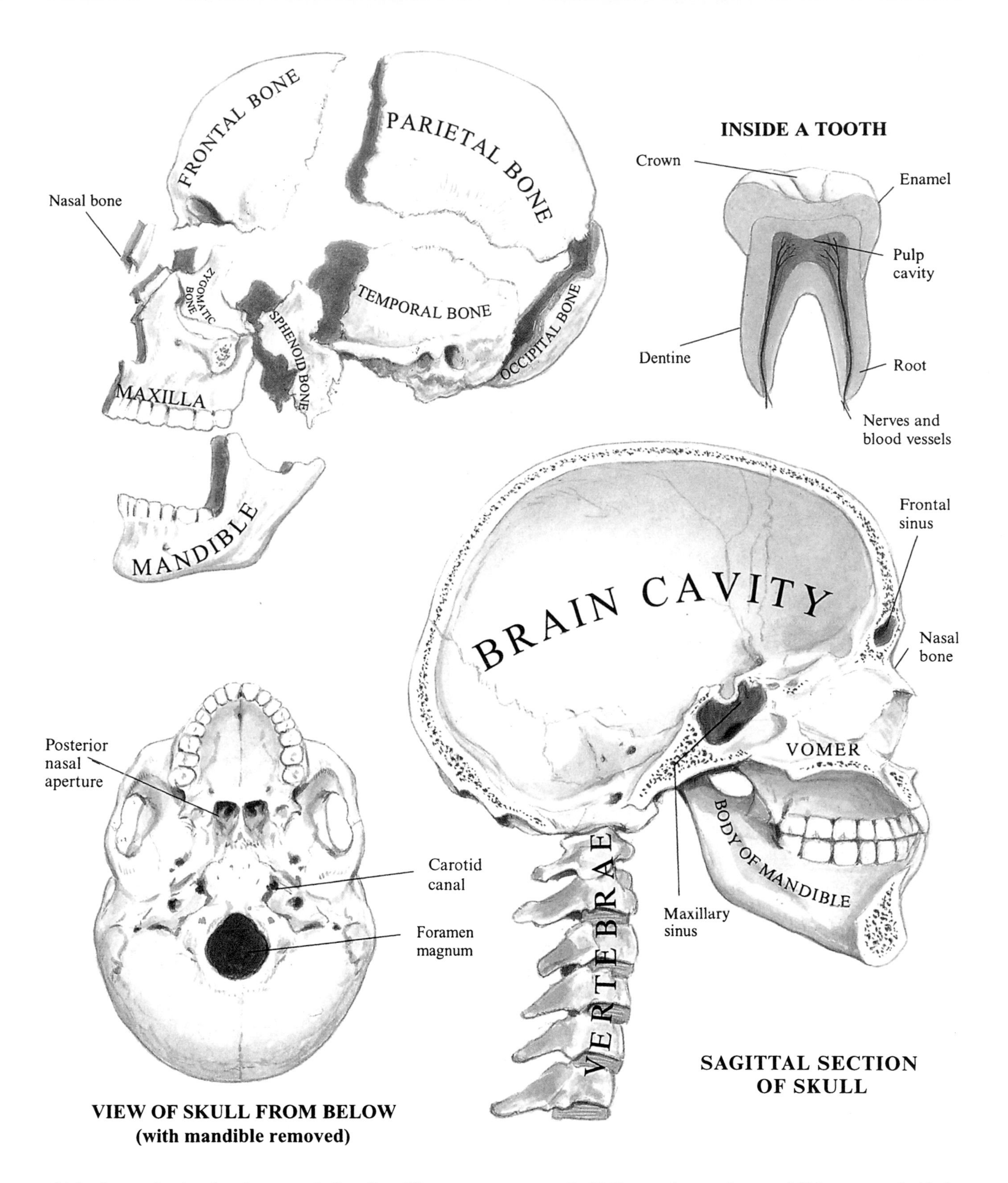

SAGITTAL SECTION OF SKULL

VIEW OF SKULL FROM BELOW (with mandible removed)

third—the nostrils—is soft and composed of cartilage. When a nose bleeds, the only way to compress the blood vessels and stop the flow is by pinching this soft part. In order to lighten the facial bones, parts are hollowed out into air sinuses (frontal and maxillary), which are common sites of infection (sinusitis). The largest cavity in the skull contains and protects the brain. The spinal column leaves the skull through the foramen magnum.

Both jaws carry an equal number of teeth. There are 20 milk (deciduous) teeth in a child. These appear in a set sequence from about six months and begin to be replaced by permanent teeth from six years. By 17–21 years the complete set of 32 has appeared with the presence of the third molar or "wisdom" tooth. Each quarter of an adult jaw has two incisors for cutting, one canine for tearing, and two premolars (two cusps) together with three molars (three or four cusps) for grinding. Every tooth has a root in a socket under the gum, a small neck at gum level and a crown, varying in shape according to its function. Crowns are composed of exceptionally hard dentine covered in shining, protective enamel. The tooth cavity contains blood vessels and nerves, damage to which gives rise to a toothache.

MUSCLES

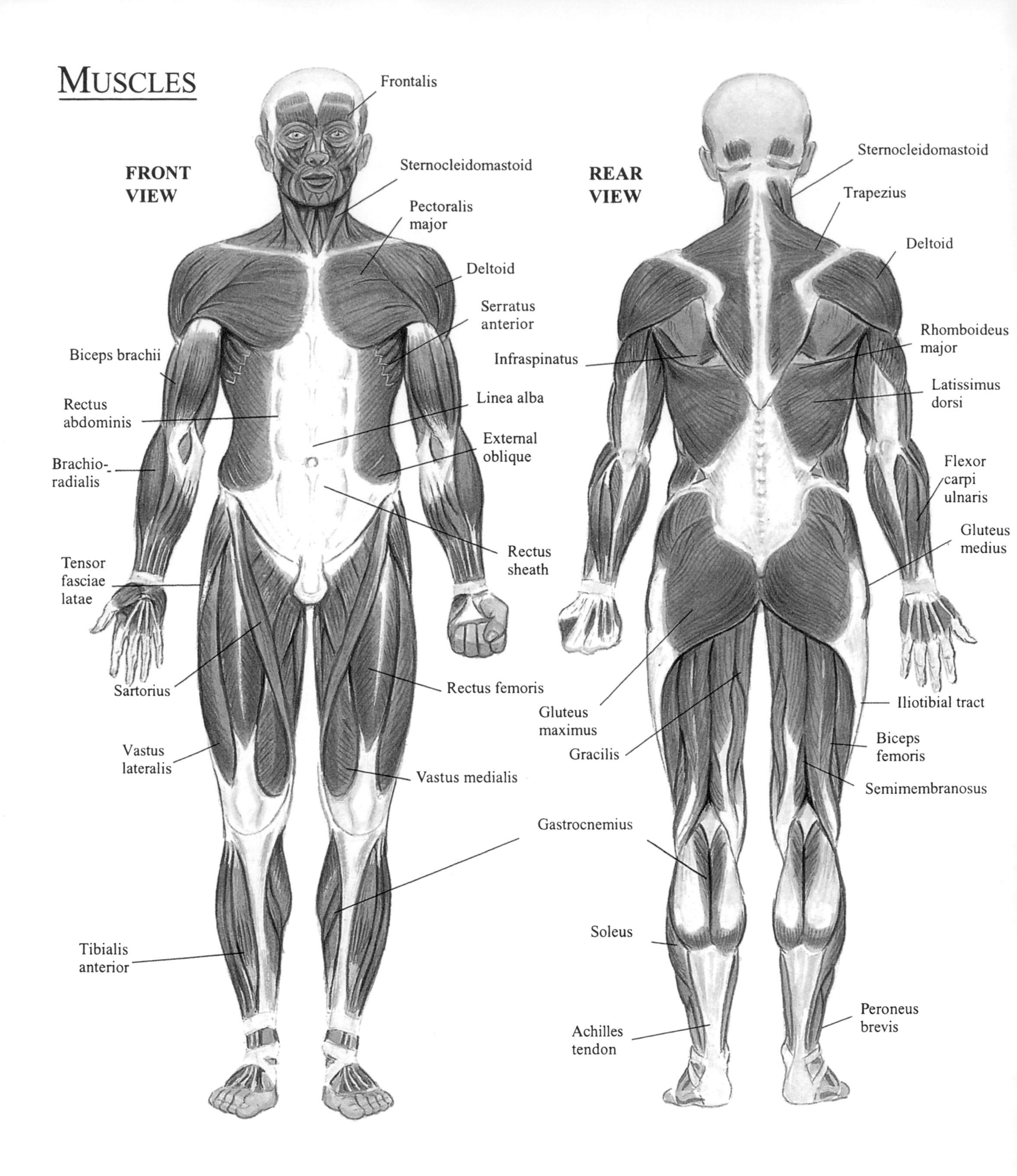

The muscles that cover the skeleton, contracting at will to make movements of bone possible, are called voluntary muscles. They also cause movements of overlying tissues, as in the face, where they create the expression of feelings, as shown by smiles and frowns. Skeletal muscle constitutes about 42 percent of a male body and 36 percent of a female body. There are three types of muscles. Striated muscle reveals patterns when viewed under a microscope. All voluntary muscles are striated. Smooth, or unstriated, muscle is found in the walls of blood vessels, the alimentary tract and the ducts of glands. Heart muscle is a specialized type of striated muscle, contracting automatically at a rate regulated by its nerve supply.

A muscle fiber can vary in size from fractions of an inch (several millimeters) to over one foot (30 centimeters). Skeletal muscle consists of bundles of fibers held together by connective tissue and enclosed in a tough fibrous sheath called fascia, which is usually connected at one or both ends to bundles of white fibrous tissue—a tendon. Tendons anchor muscles to bones and joints. In order to move bones and joints, a muscle must be fixed at its origin, which is nearer the inner aspect of the body and the upper part of a limb. The other end (the insertion) is attached to a point away from the center of the body and

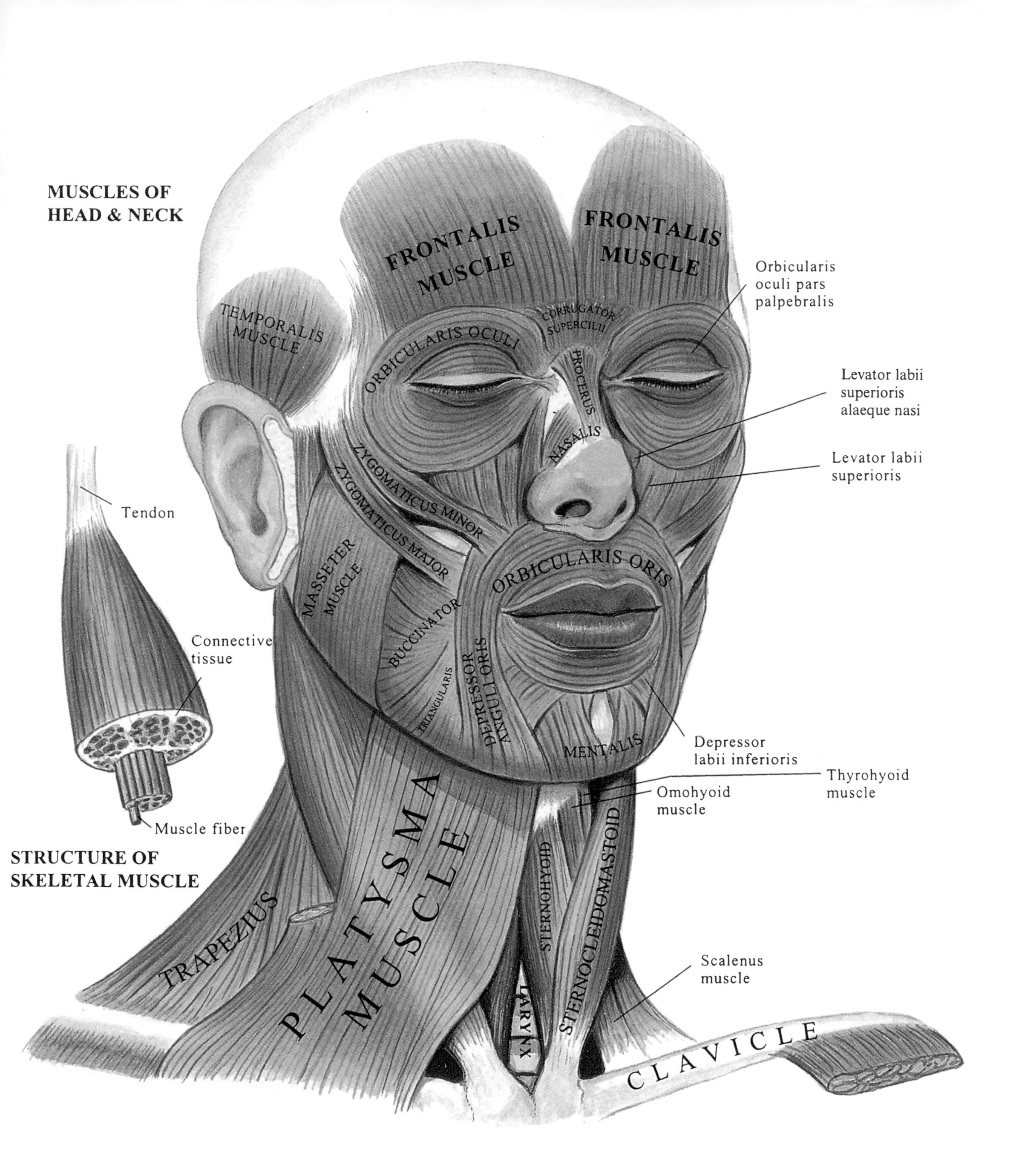

toward the end of a limb. Some attachments are complex fibrous structures. When a muscle contracts, the origin remains stationary and the insertion moves. Muscles must cross the joint that they move; some pass over more than one joint, for example, the biceps, which passes over the shoulder and the elbow. When a muscle contracts, it shortens and pulls, but cannot push. As a contraction passes away, the muscle becomes soft and longer. However, movement cannot take place unless other muscles, having the opposite action, relax—these paired muscles are called antagonists.

Energy for movements comes from glycogen, the chief carbohydrate storage material in animals, which is found in the muscles and the liver. Glycogen is broken down into carbon dioxide and water with the release of energy through a process of oxidation, in which oxygen is consumed. During leisurely activity, enough oxygen is available, but during violent exercise, there is often not enough oxygen, and lactic acid is produced instead of carbon dioxide. This buildup of lactic acid in muscle tissue gives rise to cramps and fatigue.

THE SPINE

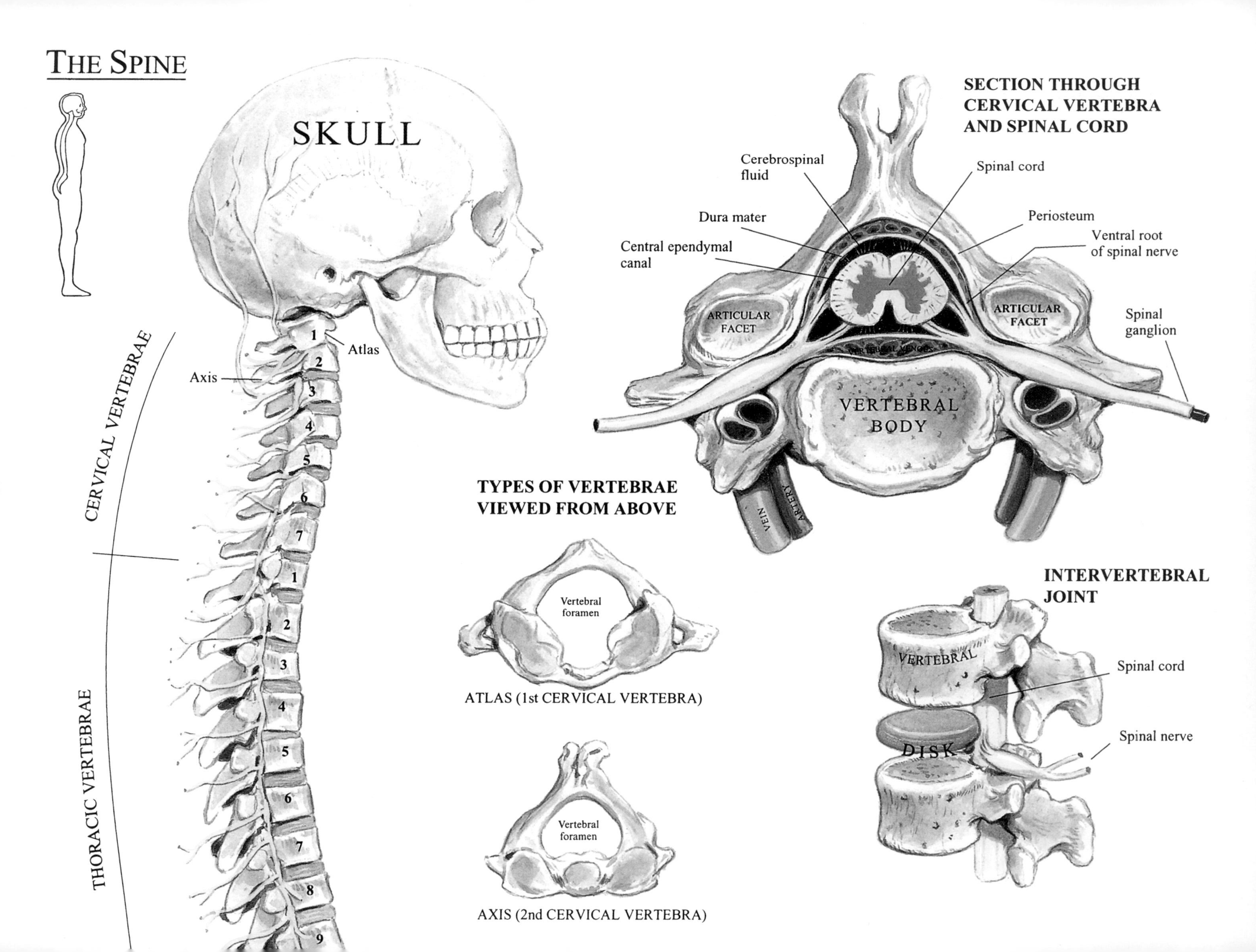

SKULL
Atlas
Axis
1
2
3
4
5
6
7
1
2
3
4
5
6
7
8
9
CERVICAL VERTEBRAE
THORACIC VERTEBRAE
SECTION THROUGH CERVICAL VERTEBRA AND SPINAL CORD
Cerebrospinal fluid
Spinal cord
Dura mater
Periosteum
Central ependymal canal
Ventral root of spinal nerve
ARTICULAR FACET
ARTICULAR FACET
Spinal ganglion
VERTEBRAL BODY
VEIN
ARTERY
TYPES OF VERTEBRAE VIEWED FROM ABOVE
Vertebral foramen
ATLAS (1st CERVICAL VERTEBRA)
Vertebral foramen
AXIS (2nd CERVICAL VERTEBRA)
INTERVERTEBRAL JOINT
VERTEBRAL
Spinal cord
DISK
Spinal nerve

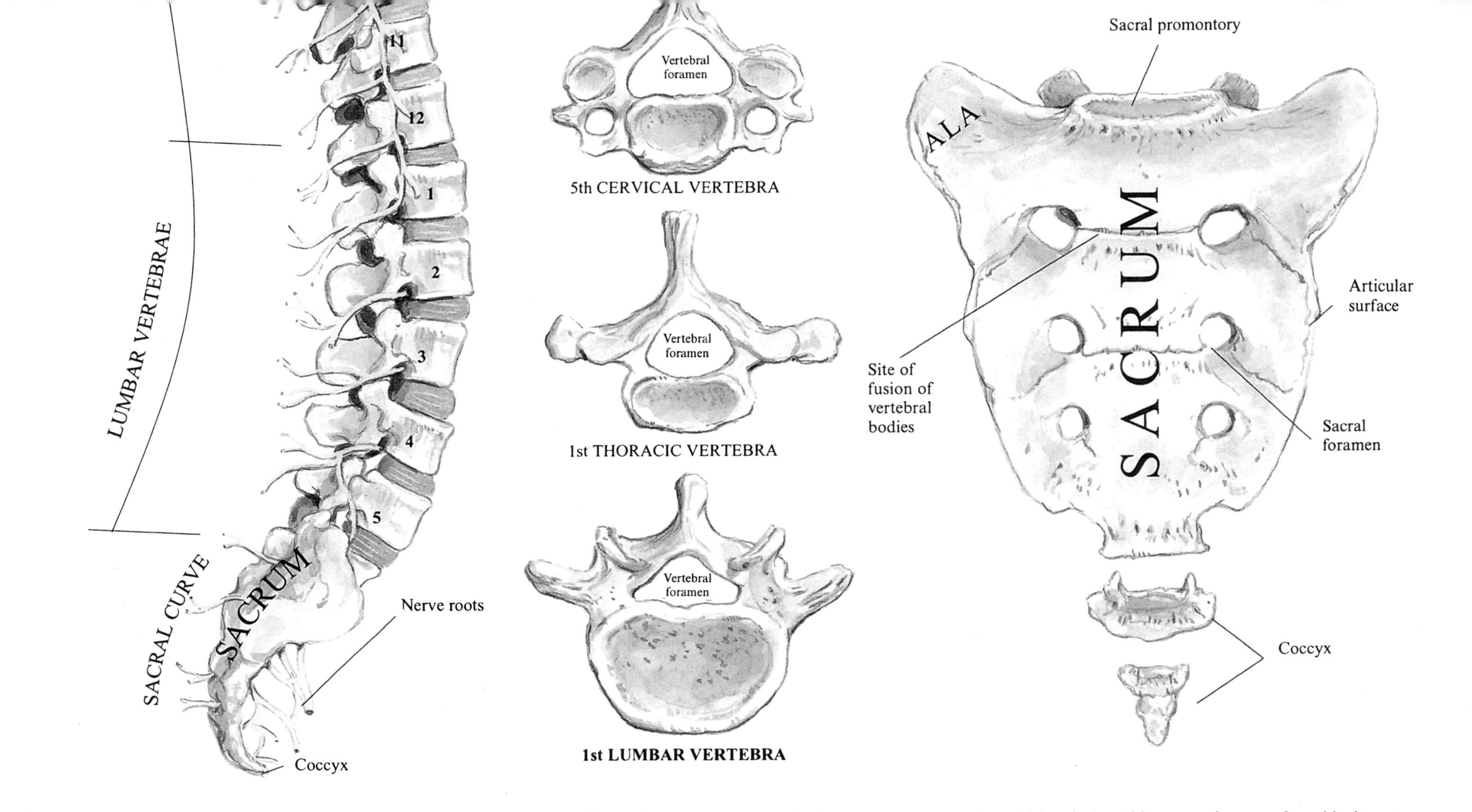

The spine, also known as the spinal or vertebral column, stretches from the base of the skull to the coccyx and is composed of bones called vertebrae. Disks of fibrocartilage, known as intervertebral disks, separate the body of each vertebra with ligaments in front and behind to keep the spinal column together in its characteristic S-shaped curve. The column is flexible and maintains the upright posture of the body. A typical vertebra has a thick, strong body, facing frontward with an arch of bone pointing backward and enclosing the spinal cord in the space called the vertebral foramen. The seven cervical vertebrae in the neck are the smallest, the first two allow rotation of the skull. They are so specialized in shape that they have their own names—atlas and axis. The 12 thoracic vertebrae increase in size from top to bottom and have long spines and extra facets to articulate with the ribs. The top eight join with two pairs of ribs—its own and the one below. The lower four join with their own ribs. Because they keep the upper and lower parts of the body together, the five lumbar vertebrae have very thick, strong bodies with heavy spinous processes for the attachment of muscles. The sacrum, which gives stability to the pelvis between the hips, is composed of five vertebrae fused to form a triangular bone. The coccyx, also triangular, is composed of four rudimentary vertebrae which articulate with, or sometimes even fuse with, the sacrum.

The spinal cord extends from the base of the skull to the first lumbar vertebra—an average distance of 17½ inches (45 centimeters). It branches into pairs of nerves, each having an anterior (front) motor root, which initiates activity, and a posterior sensory root to the back of the cord. The two roots join to form spinal nerves. There are 31 pairs made up of eight to the neck or cervical region, 12 to the chest (thoracic nerves), five to the lower back (lumbar region), five to the sacrum (sacral nerves) and one to the coccyx. Some spinal nerves join up to form a plexus: cervical 1–4 go to the neck and shoulder with the phrenic nerve serving the diaphragm; cervical 5 to thoracic 1 form the brachial plexus which becomes the radial, median and ulnar nerves in the forearm. Lumbar 1–4 nerves form the femoral nerve, which runs down the front of the thigh, and lumbar nerves 4 and 5 and sacral 1–3 form the sciatic nerve, the largest in the body, extending down the back of the leg, dividing above the knee into peroneal and tibial nerves. The sciatic nerve is commonly damaged in back injuries.

THE BRAIN

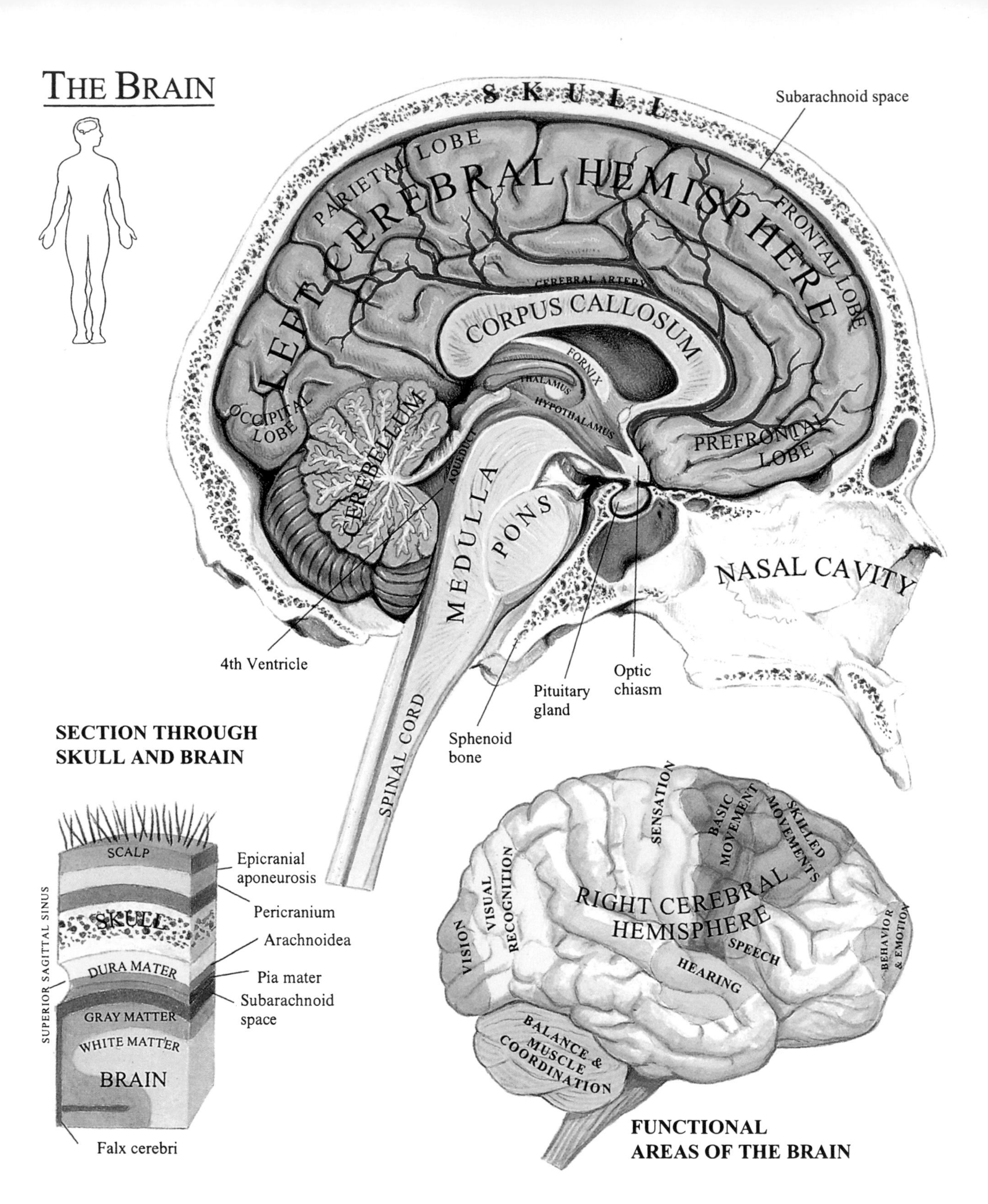

The brain is in three parts—fore, mid and hind—protected by three membranes: the dura mater, arachnoidea and pia mater, collectively called the meninges. The largest part of the forebrain is the cerebrum, divided into two cerebral hemispheres by a huge fissure. Each half—left and right—has sensory areas for receiving messages and motor areas for sending out messages, normally to the opposite side of the body. The outer layer of the cerebrum is the cerebral cortex, composed of nerve cells (gray matter). The tissue below appears white because it contains nerve fibers. The hemispheres are joined by a band of nerve fibers called the corpus callosum. The rest of the forebrain contains the egg-shaped thalamus, which integrates the sensory, motor and emotional processes of the body; the hypothalamus, which regulates the autonomic system (nerves controlling the heart, glands and smooth muscles) and the pituitary gland below it by secreting hormones into the bloodstream. The midbrain is small—about ¾ inch (two centimeters) long—connecting the forebrain with the hindbrain (principally the cerebellum), which coordinates the accuracy of muscle activity and maintains the posture of the body. The brain stem consists of the pons, carrying communication fibers to and from the cerebellum and the medulla oblongata. The medulla oblongata contains nerve centers essential to life, which control the heart and respiration. The twelve pairs of cranial nerves supply the organs of the head and neck, extending via the tenth (vagus) nerve to the chest, heart and abdomen.

NERVOUS SYSTEM

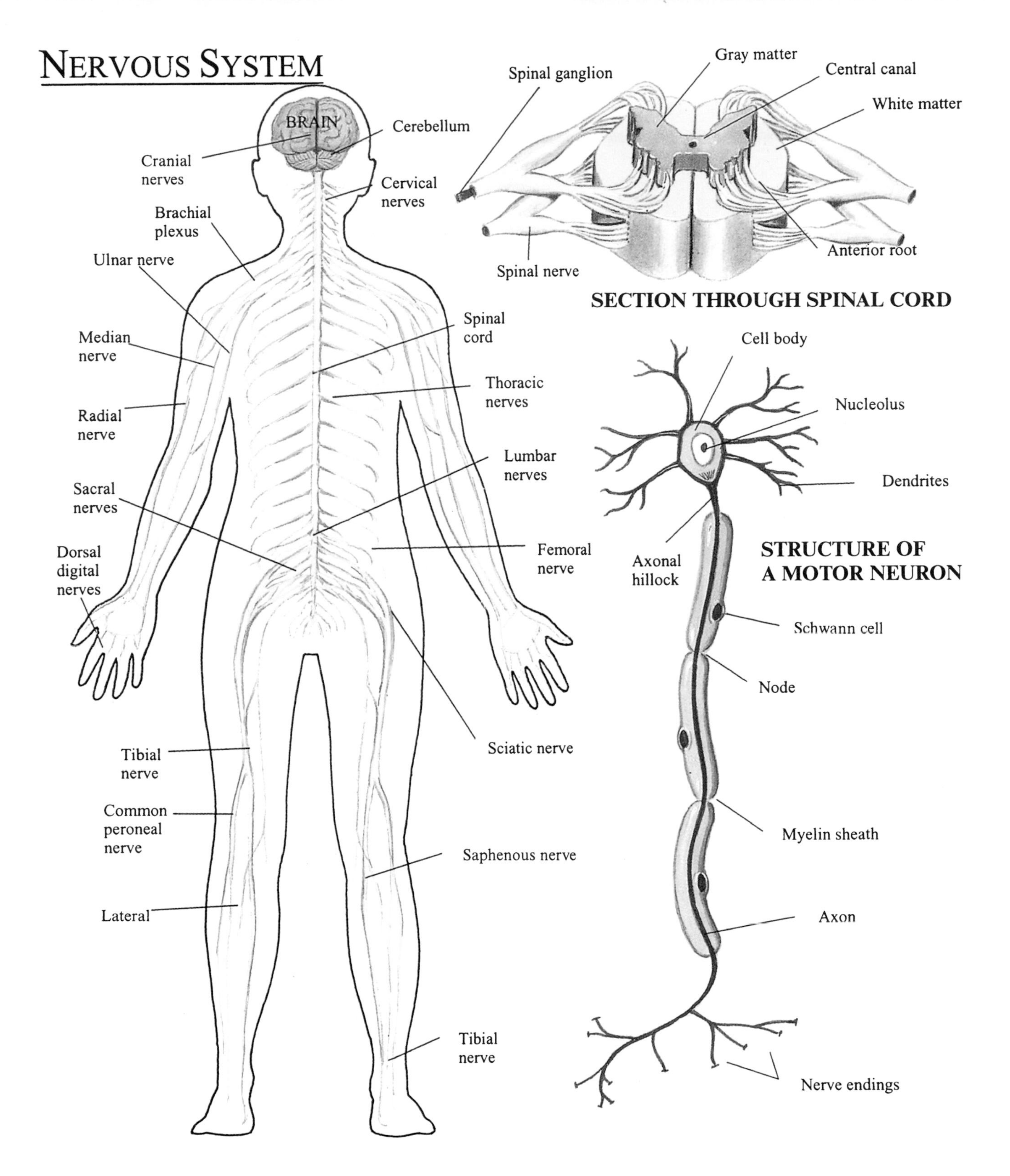

The nervous system is the communications network of the body. It has two parts—the central nervous system, consisting of the brain and spinal cord, and the peripheral nervous system, radiating from the spinal cord to muscles, internal organs and skin. Sensations from the environment outside the body, internal organs and other tissues are interpreted and the appropriate reaction is registered by the brain. Nerve fibers form the bulk of the brain, spinal cord and peripheral nerves. Every nerve fiber consists of a thread-like axon, which may range from a few centimeters to many centimeters in length. The fiber may or may not be covered by a fatty myelin sheath, produced by Schwann cells. Nerve fibers are classified as myelinated or nonmyelinated. There are gaps in the myelin sheath called nodes of Ranvier where nutrients can be absorbed and waste products excreted. Disease can occur when the myelin covering is lost. Myelinated fibers form the white matter of the brain and spinal cord, whereas nerve cell bodies comprise the gray matter. The basic unit of the nervous system is a neuron, consisting of a nerve cell with several branching processes (dendrites) which receive nerve impulses (waves of changing electrical energy). These pass down myelinated fibers faster—at over 328 feet (100 meters) per second—than through nonmyelinated fibers. Impulses are passed from the end of one fiber to another via a synapse or to a muscle fiber at a neuromuscular junction.

THE SKIN & HAIR

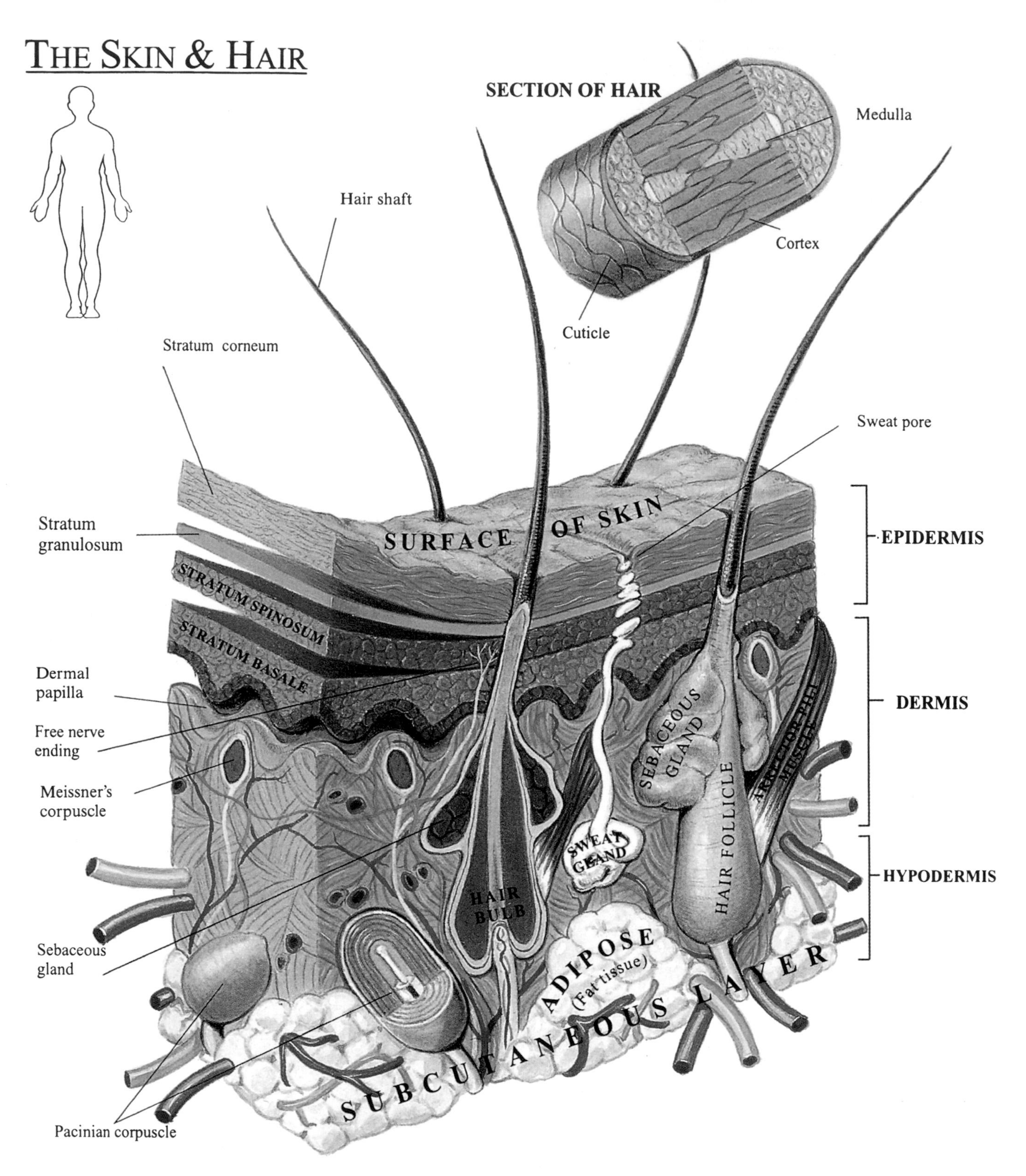

The skin—strong enough to withstand everyday knocks, yet supple enough to allow movement and varying in thickness from 1/16 inch (5 millimeters) on the soles of the feet to less than 1/32 inch (0.5 millimeters) on the eyelids—envelops the whole body. Accounting for 15 percent of adult body weight, skin is the body's largest organ. Its functions include sensing changes in the environment, controlling body temperature, acting as a waterproof barrier or a screen against damaging radiation from the sun and protecting underlying tissues from infection. Skin has two principal layers—the dermis and epidermis. The underlying dermis contains fat cells, blood and lymphatic vessels, sweat and sebaceous glands, elastic fibers and Meissner's corpuscles, which detect touch and are more plentiful on tips of fingers, toes and lips. Oil from sebaceous glands keeps the skin surface supple, maintaining its texture and waterproof properties. The epidermis consists of layer upon layer of cells in varying stages of destruction—flat and without nuclei (stratum corneum), granular (stratum granulosum) and those with spines (stratum spinosum)—all derived from the base layer (stratum basale). The outer layers of skin, containing keratin, a tough protein also present in nails and hair, are continually shed and replaced by those below. The skin is colored by various pigments in the epidermis, including melanin, which changes with exposure to sunlight, giving a tan. Sweat glands help to keep body temperature constant by secreting sweat, which evaporates to cool the skin. Hair shafts grow from follicles. After about 1000 days a hair shaft is shed, to be replaced. About 100 are shed each day.

NOSE, MOUTH & THROAT

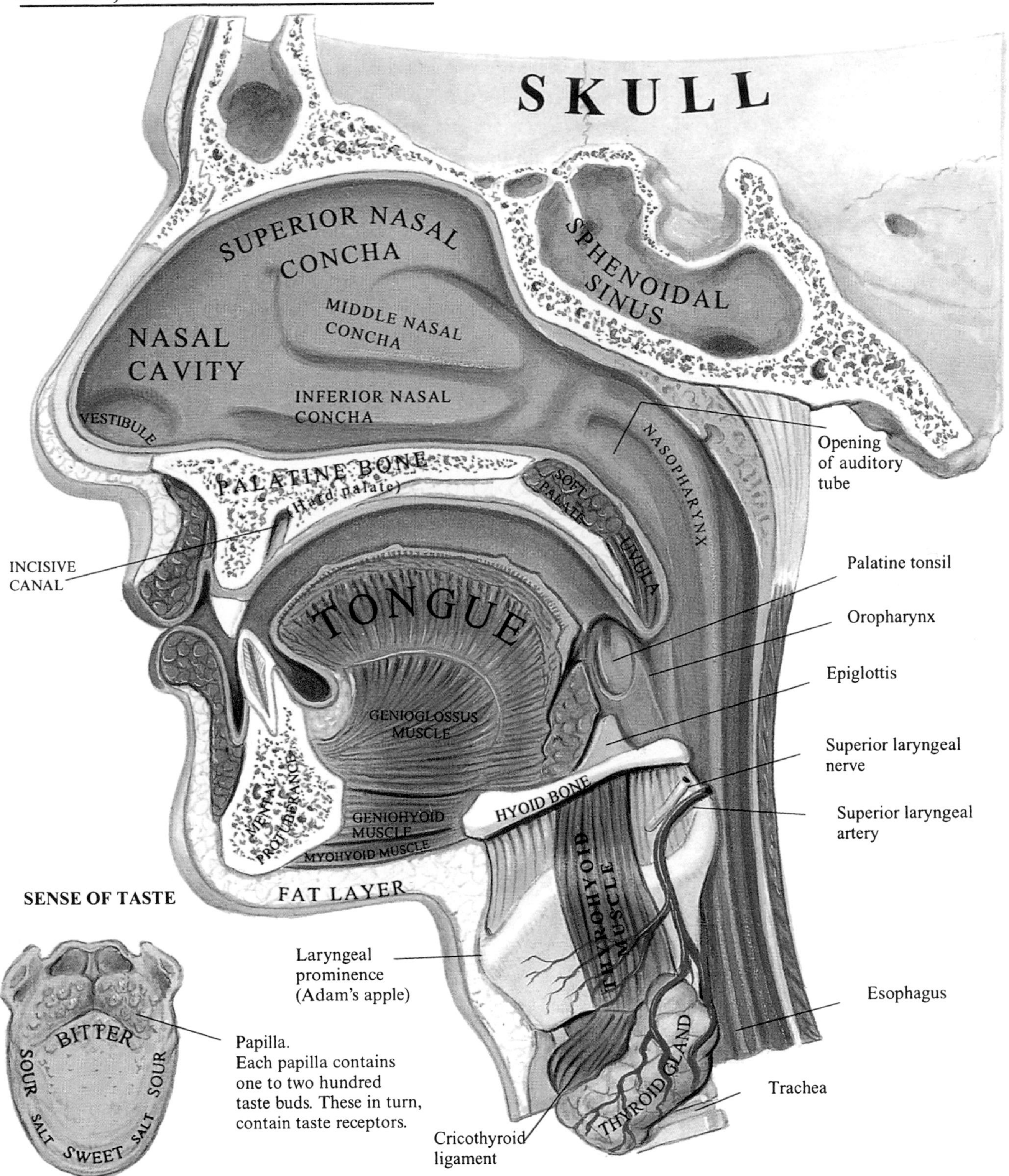

The nose, mouth and throat are concerned with breathing, smelling, speaking and eating. The taste of food is derived from the senses of smell and taste. Without smell, only sweet, sour, salt and bitter flavors can be appreciated by the taste buds, located on the surface of the muscular tongue, which is rooted on the hyoid bone. The tongue is used in eating, speaking and swallowing.

Air, drawn through the nose into a cavity lined by a moist membrane, is humidified and warmed. Hairs in the lining trap dust, while nerve endings sense smells and transmit this information to the brain via the olfactory nerve. Nasal folds, three conchae and the adjoining nasal sinuses increase the area of contact. Secretions are drained backwards into the pharynx and swallowed. Excess mucus production, as happens in cold weather or during an infection (a "cold"), produces a "running nose." The pharynx can be seen by opening the mouth, protruding the tongue and saying "Ah." This reveals the soft palate, ending in a midline fold (uvula) which separates two masses of lymphoid tissue (the tonsils). During swallowing, the entrance to the airway (trachea) is closed by a movable cartilaginous flap (epiglottis). This prevents food from "going the wrong way" into the lungs. Below the hyoid bone, the trachea becomes the voice box (larynx), where the vocal cords lengthen and shorten to produce sounds. Cartilage around the larynx is fused to form the Adam's apple, more prominent in men than in women.

THE EAR

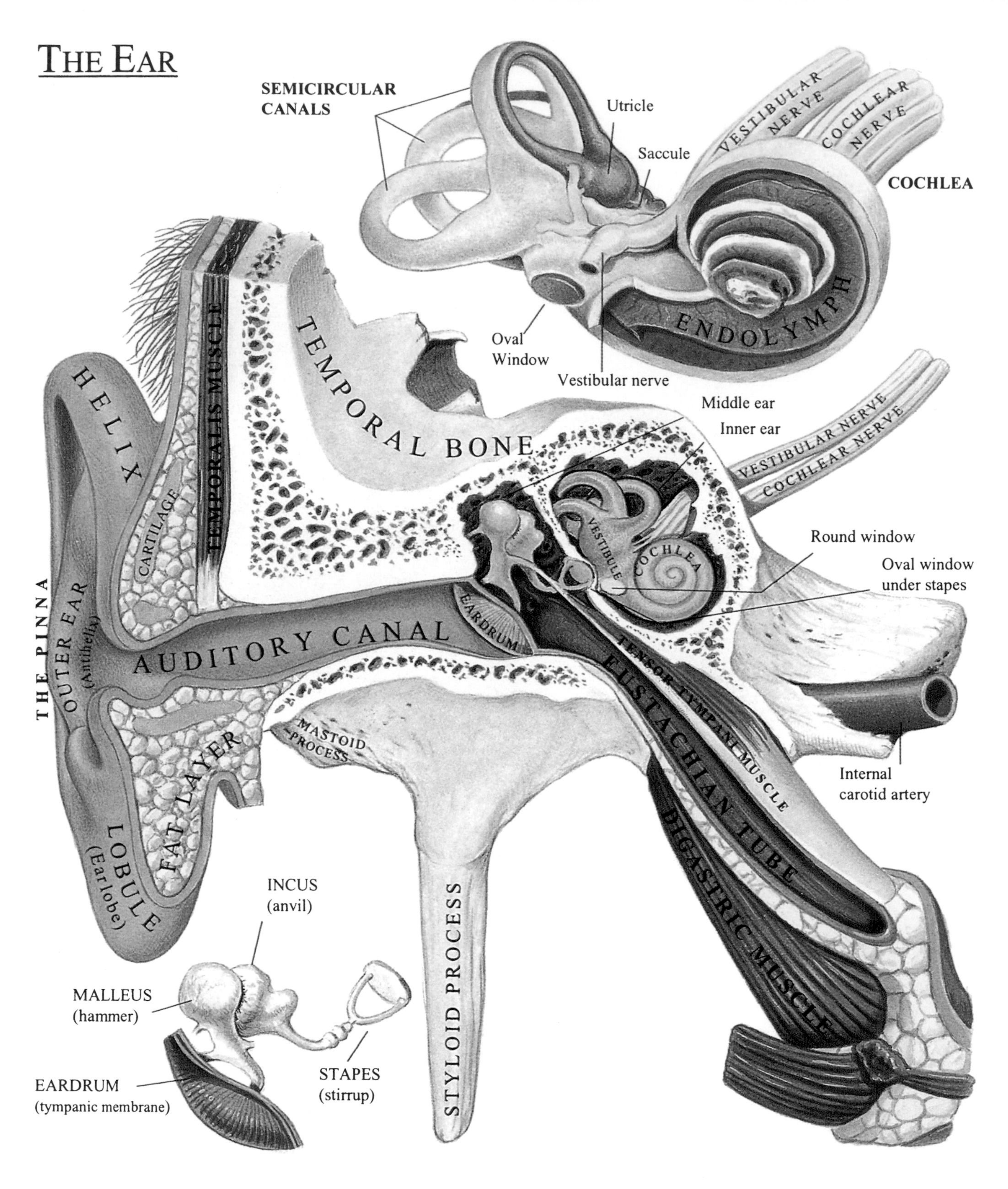

The organ of hearing and balance, the ear, is divided into outer, middle and inner parts. The visible, outer ear (pinna) funnels sound waves down the auditory canal onto the tympanic membrane stretched across the tube like a drum, separating it from the middle ear. Sound waves cause the eardrum to vibrate and move three little bones, the malleus, incus and stapes, otherwise known as the hammer, anvil and stirrup. The malleus is attached to the inner side of the eardrum and to the incus, which is also connected (by the stapes, the smallest bone in the body) to another membrane, called the oval window. A narrow canal, the Eustachian tube, extends from this part of the middle ear to the back of the throat. It equalizes the air pressure on both sides of the eardrum. The three bones amplify sound waves and enable them to pass through the oval window into the inner ear, where they make the fluid in the cochlea, a coiled tube, vibrate. The cochlea is lined with sensitive hairs that set up electrical impulses when they are disturbed. The cochlear nerve conducts these impulses to the brain. The organs of balance, the utricle, saccule and three semicircular canals, are also fluid-filled, lined with hairs and connected to the cochlea. When the head is turned or tilted, the fluid moves, activating the hairs to produce electrical impulses that are transmitted by the vestibular nerve, informing the brain of the changes in position.

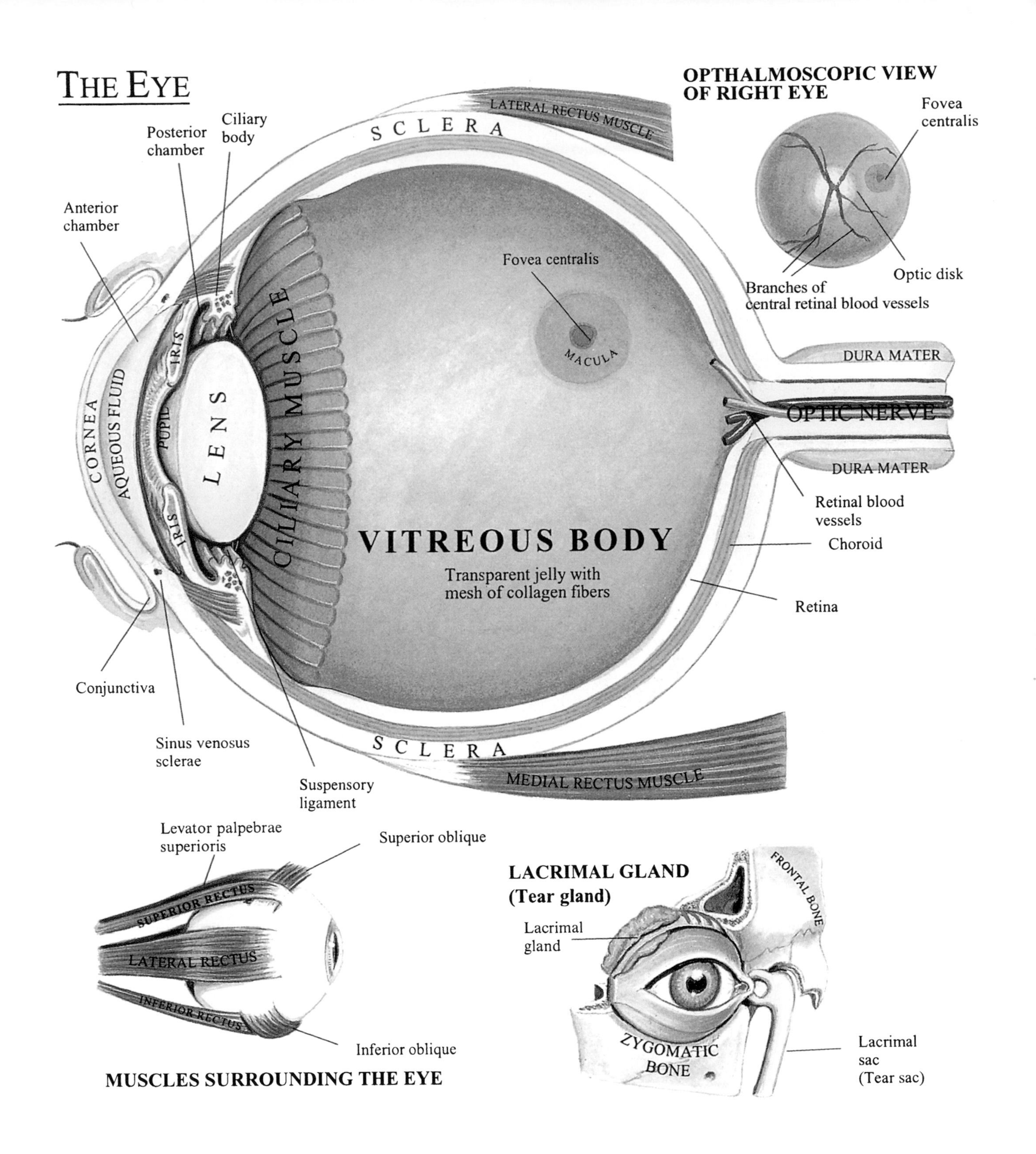

The two eyes transmit color images of the outside world to the brain. Each eye is spherical and enclosed by three layers—sclera, choroid and retina—and cushioned in fat lining a bony cradle (orbit). The tough, outer coat (sclera) forms "the white of the eye," with a clear window (cornea), covered by a moist membrane (conjunctiva) continuous with that of the eyelids, which is lubricated by tears from the tear ducts (lacrimal glands). The pigmented, dark brown middle layer of the eyeball (choroid) supplies blood and nutrients to the eye. The lens, which can be altered in shape by its attached muscles (focusing by accommodation), is suspended from the circular, glandular ciliary body, which secretes transparent aqueous fluid. The lens and cornea bend light rays to focus on the light-sensitive lining (retina), which contains special cells (rods and cones). Cones, concerned with detailed color vision, are concentrated near its center (macula). Rods are more numerous around the periphery and work in poor light. In front of the lens, a pigmented ring of muscle (iris) controls the amount of light entering the eye by widening or narrowing the pupil. Nerve fibers from the rods and cones pass over the inside of the retina to the optic nerve (optic disk). Transparent fluid (vitreous body) fills the space behind the lens. There are six muscles in the orbit that control eye movements and weakness in one or more muscles can cause an eye to deviate—called a strabismus.

THE HAND & ARM

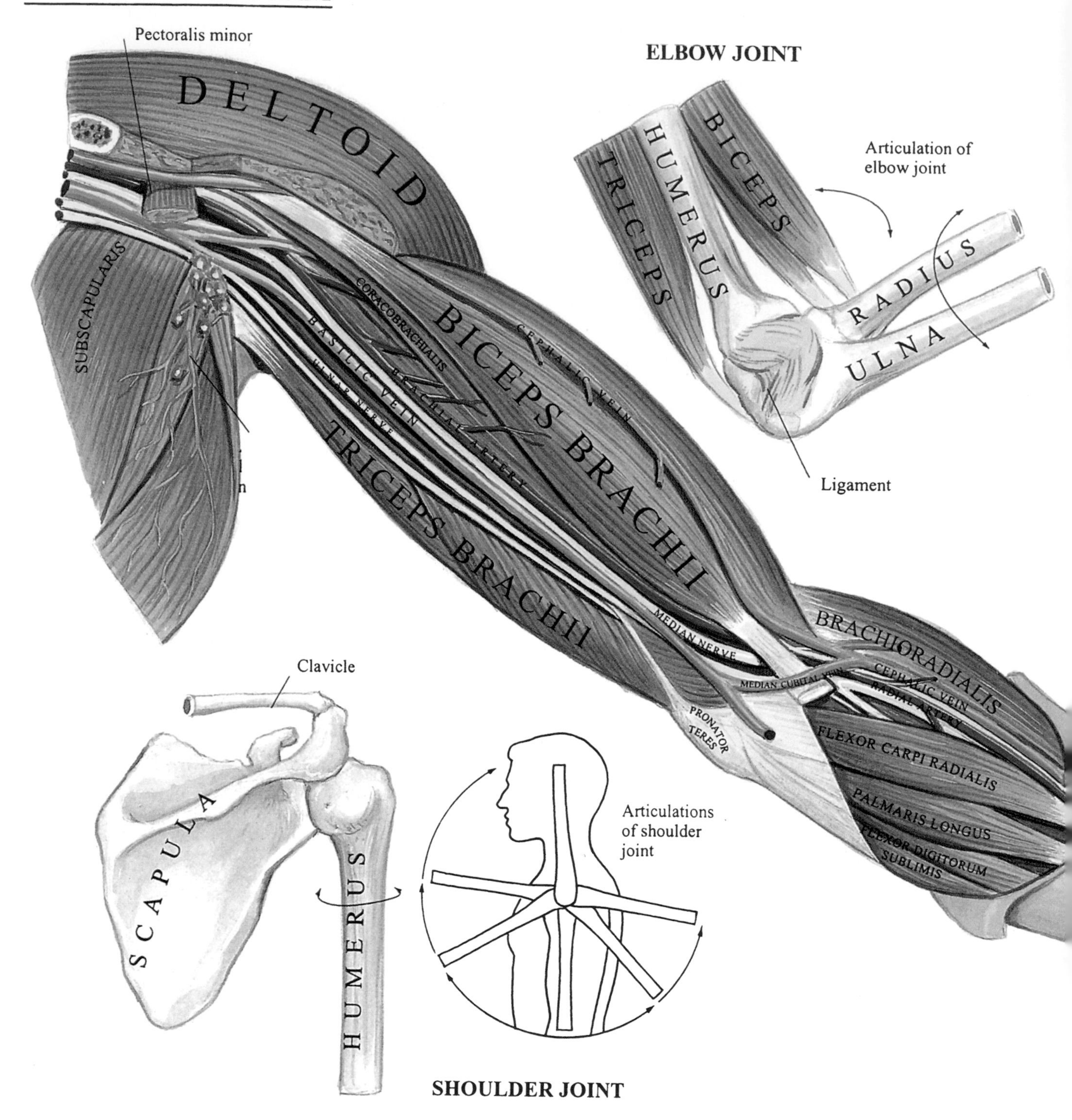

The hand and arm make up the upper limb, capable of many functions and fine movements because of its specialized joints and muscles. The shoulder joint is the most mobile in the body, a shallow cup and loose ligaments holding the scapula (shoulder blade) to the humerus of the upper arm. The scapula is triangular and mobile, fixed only by muscles to the back of the chest. The large deltoid muscle covers the shoulder joint above; below it lies the biceps muscle with the triceps on the underside of the upper arm. The clavicle (collarbone) produces the only bony stability by articulating with the sternum (breastbone). Nevertheless, the shoulder joint, which can move the arm in a complete circle, is easily wrenched out of its socket. Such a derangement is called a dislocation. By bending the elbow, the biceps muscle can be made to stand out below the shoulder joint. Grip the hand tightly and the forearm muscles become prominent. In order to do this, muscles with opposite actions relax.

There are two bones in the forearm—the radius and ulna—which, with their muscles, can rotate the hand from palm down to palm up. Eight small, irregularly shaped bones make up the wrist joint. The

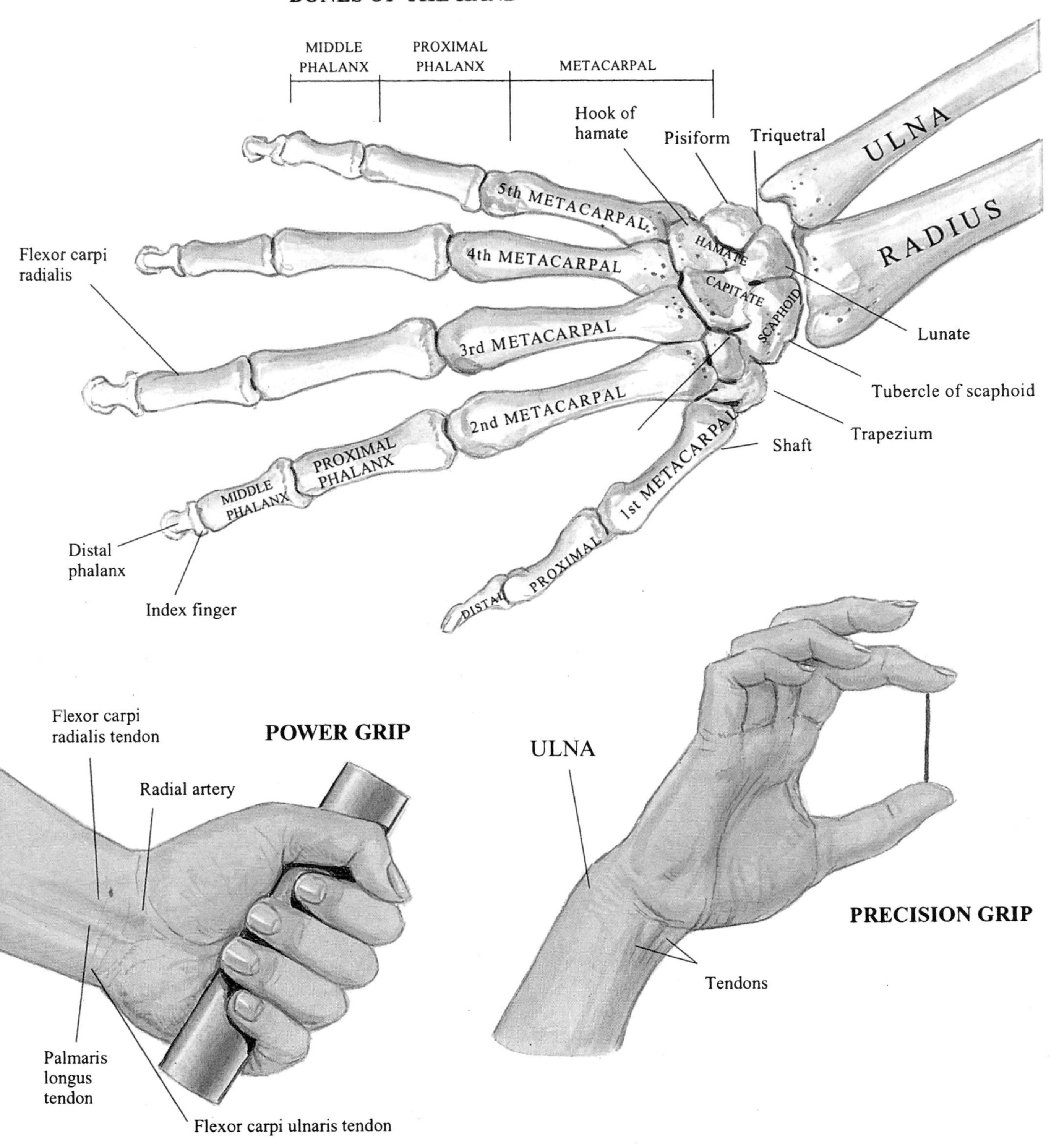

bones in the hand and wrist have many specialized muscles attached to them, permitting great mobility. Thus, the hand is capable of such very fine movements as writing, painting, making tiny models or repairing a car. The fingers can be brought together to touch one another (called approximation), or hold an object in a precision grip. Closing the fingers to the palm of the hand produces a fist. Muscles in the forearm cross into the hand as tendons on both sides of the wrist to minimize bulk and enable movements to be made upward and downward. The bones of the palm of the hand are called metacarpals; the knuckles correspond to their ends and the joints with the fingers. The digits (phalanges) are divided into a thumb and four fingers, in turn designated the first or index finger, middle, ring finger and little finger. Each finger consists of three bones—a near (proximal) phalanx, middle and that farthest away (distal). The saddle joint of the base of the thumb enables it to move across the palm (opposition). The skin over the last bone (distal phalanx) in each finger produces a nail, a horny plate that grows throughout life.

THE LEG & FOOT

HIP JOINT

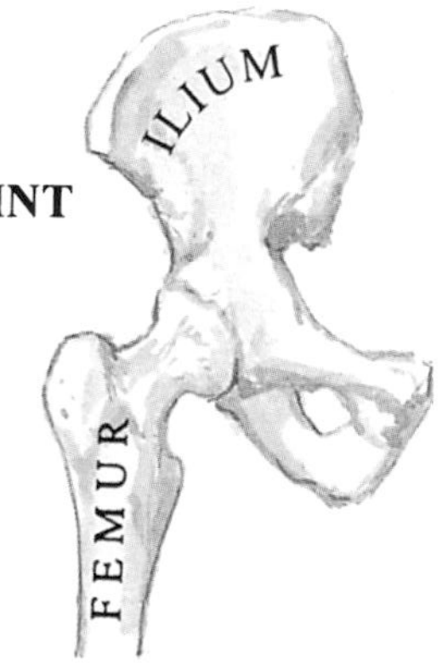

ILIUM
FEMUR

KNEE JOINT
WITH LIGAMENTS

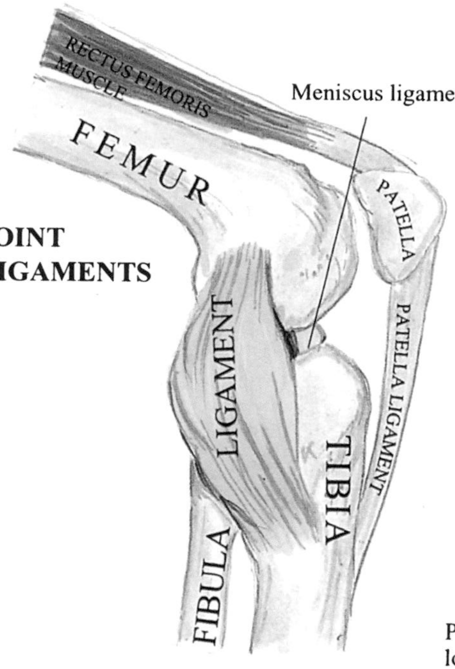

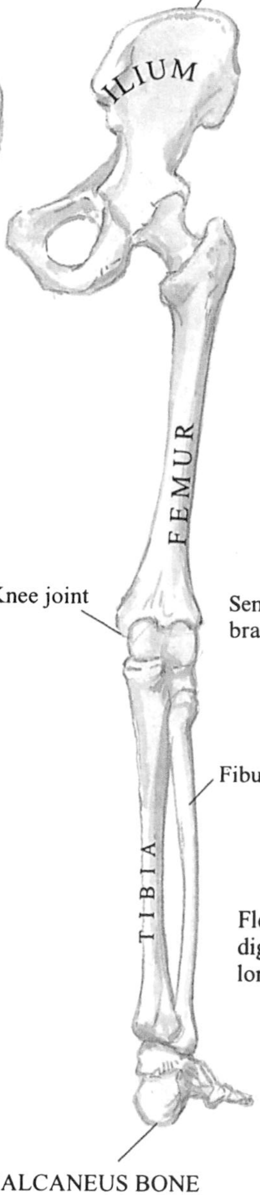

RECTUS FEMORIS MUSCLE
FEMUR
Meniscus ligament
PATELLA
PATELLA LIGAMENT
LIGAMENT
TIBIA
FIBULA

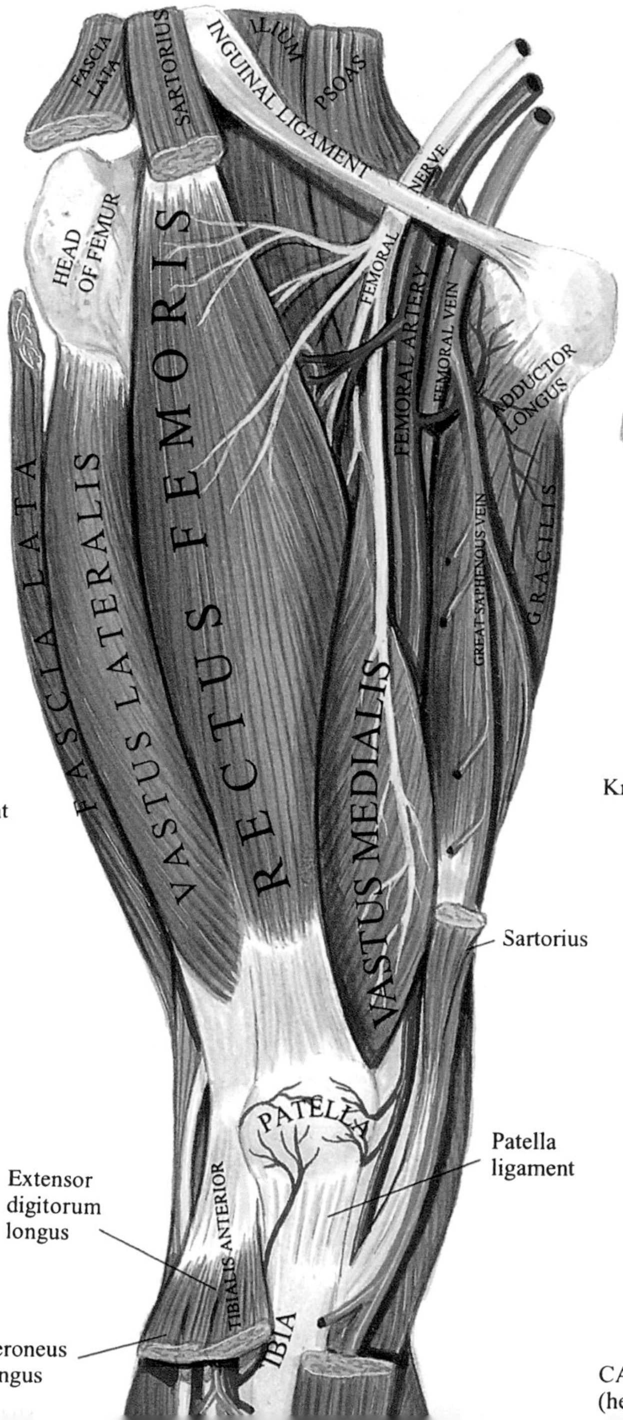

FASCIA LATA
SARTORIUS
INGUINAL LIGAMENT
ILIUM
PSOAS
HEAD OF FEMUR
FEMORAL NERVE
FEMORAL ARTERY
FEMORAL VEIN
ADDUCTOR LONGUS
GRACILIS
GREAT SAPHENOUS VEIN
FASCIA LATA
VASTUS LATERALIS
RECTUS FEMORIS
VASTUS MEDIALIS
Sartorius
PATELLA
Patella ligament
Extensor digitorum longus
TIBIALIS ANTERIOR
TIBIA
Peroneus longus

Crest of ilium
ILIUM
FEMUR
Knee joint
Fibula
TIBIA
CALCANEUS BONE (heel bone)

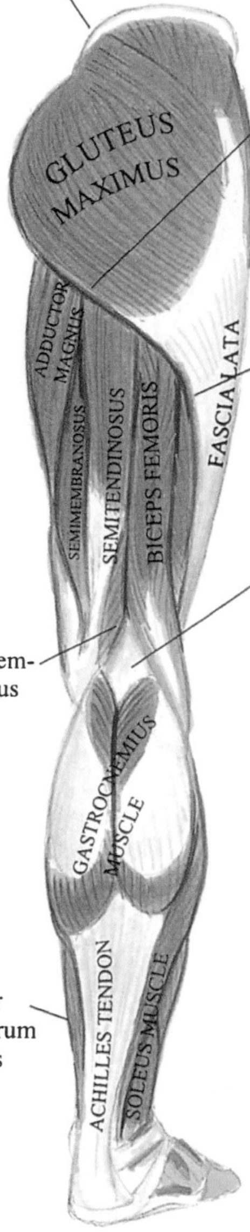

GLUTEUS MAXIMUS
Adductor magnus
ADDUCTOR MAGNUS
SEMIMEMBRANOSUS
SEMITENDINOSUS
BICEPS FEMORIS
FASCIA LATA
Vastus lateralis
Popliteal fossa
Semimembranosus
GASTROCNEMIUS MUSCLE
Flexor digitorum longus
ACHILLES TENDON
SOLEUS MUSCLE

TENDONS AND TENDON SHEATHS OF THE FOOT

BONES OF THE FOOT

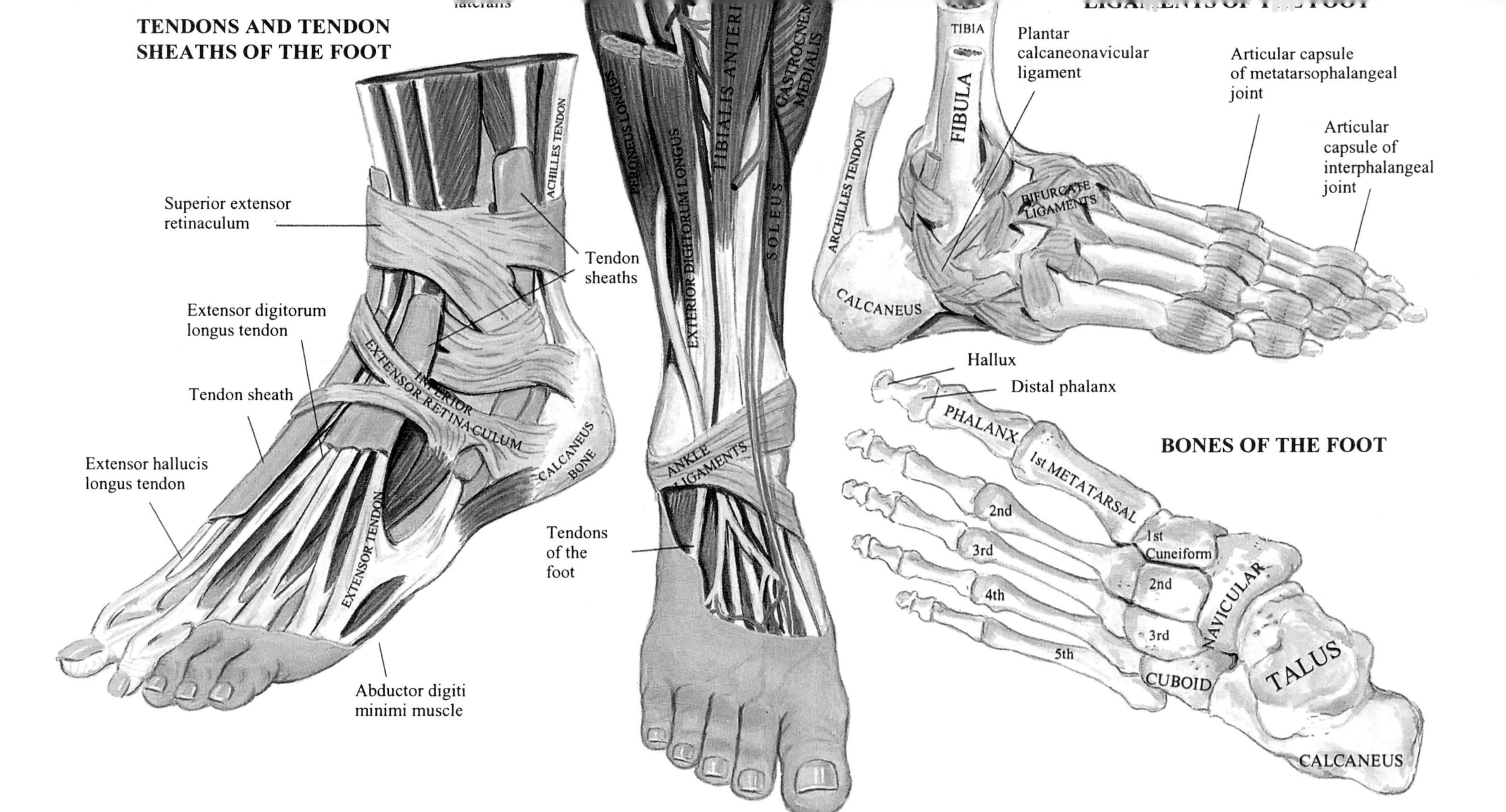

The leg and foot form the lower limb, which supports and carries the weight of the whole body while standing, walking, leaping and running. During these movements weight is transferred from one leg to the other, sometimes rapidly, sometimes forcefully. Strength and stability are essential to withstand the stresses involved. The gluteus maximus is the heaviest muscle of the body, stretching from the pelvis to mid-thigh, forming and shaping the buttocks. The femur is the longest—often up to one third of the body length—and the strongest bone in the body. The hip joint is the deepest ball-and-socket joint and very stable. However, fractures occur at the neck of the femur, especially in old people. The knee joint is the biggest and most complicated in the body, functioning as a modified hinge joint which, with its internal and external ligaments, has massive stability.

Powerful muscles act on each part of the leg. At the front of the leg, four—the rectus femoris, vastus intermedius, vastus lateralis and vastus medialis—combine to form the quadriceps femoris, in which the patella, the largest sesamoid bone in the body, is located before its insertion into the tibia to add extra stability to the joint. At the back of the knee, the division of the three flexor tendons (semimembranosus, semitendinosus and biceps femoris, known as the hamstrings) from the thigh forms a muscle-free zone, known as the popliteal fossa. These muscles are very strong and bend the knee for such activities as walking, jumping and climbing.

The gastrocnemius muscle forms the bulge of the calf and, combined with the underlying soleus muscle, ends in the Achilles tendon—the most important and powerful in the body—inserted into the calcaneus bone of the heel. If this tendon ruptures, walking becomes impossible. Muscles from the lower leg cross the ankle into the foot as tendons, protected by fluid-filled sheaths. The ankle joint is at the junction of the tibia and fibula with the tarsal bones; if it is injured, the lower tip of the tibia may break.

The foot has seven tarsal and five metatarsal bones. The great or big toe has two phalanges, as big as the three in the other toes. Like the hand, each distal phalanx has a nail.

Thanks to the design of the tarsal bones and surrounding associated ligaments, the foot is drawn up into longitudinal and transverse arches that distribute the weight of the body evenly and act as a lever for jumping. Most of the foot muscles are present as tendons but four layers of muscle on the sole of the foot help support the arches and act as padding to absorb the pounding that occurs in walking and running.

Upper Body

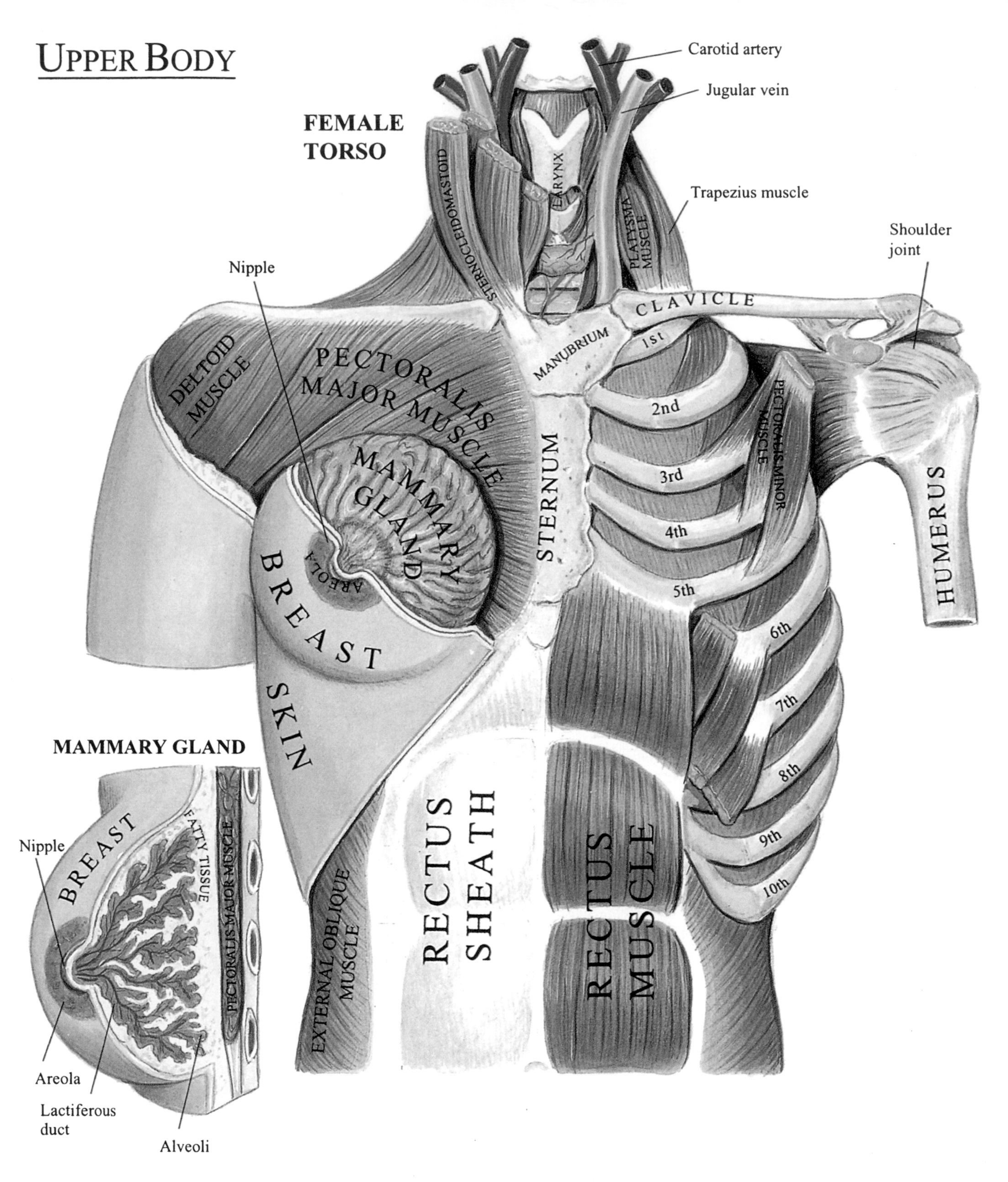

The upper body differs externally in adult men and women due principally to the presence, in women, of two enlarged breasts, which enable a mother to suckle her baby. Nipples surrounded by pink skin (areola) are present in both sexes, but at puberty, under the influence of female hormones, a girl's breasts enlarge between ribs 2 and 6, forming a mass of glandular cells (alveoli), lying on the pectoralis major muscle and divided by fat and fibrous tissue into 15 to 20 lobes. Each lobe has its own lactiferous duct opening at the nipple. During pregnancy the breasts swell, the areola becomes brown and, as birth approaches, watery fluid (colostrum), which becomes milk by the third day after delivery, is produced.

The main features of a man's torso include wider shoulders and a more prominent voice box (larynx or Adam's apple), which produces a deep voice. The chest is frequently hairy. The overall shape of a man's trunk is triangular, with a less obvious waist—the external dividing line between upper and lower body, created by the gap between rib cage and pelvis. At the waist, the only support for the body is the spinal column behind and, in front, two rectus abdominis muscles—one on either side of the midline—enclosed in a strong sheath. These powerful, straplike muscles form the abdominal wall, which protects the contents of the cavity.

Both sexes share the basic bone structures and other muscles, such as those that support the head (trapezius and sternocleidomastoid) and move the arms (deltoid).

BODY ORGANS

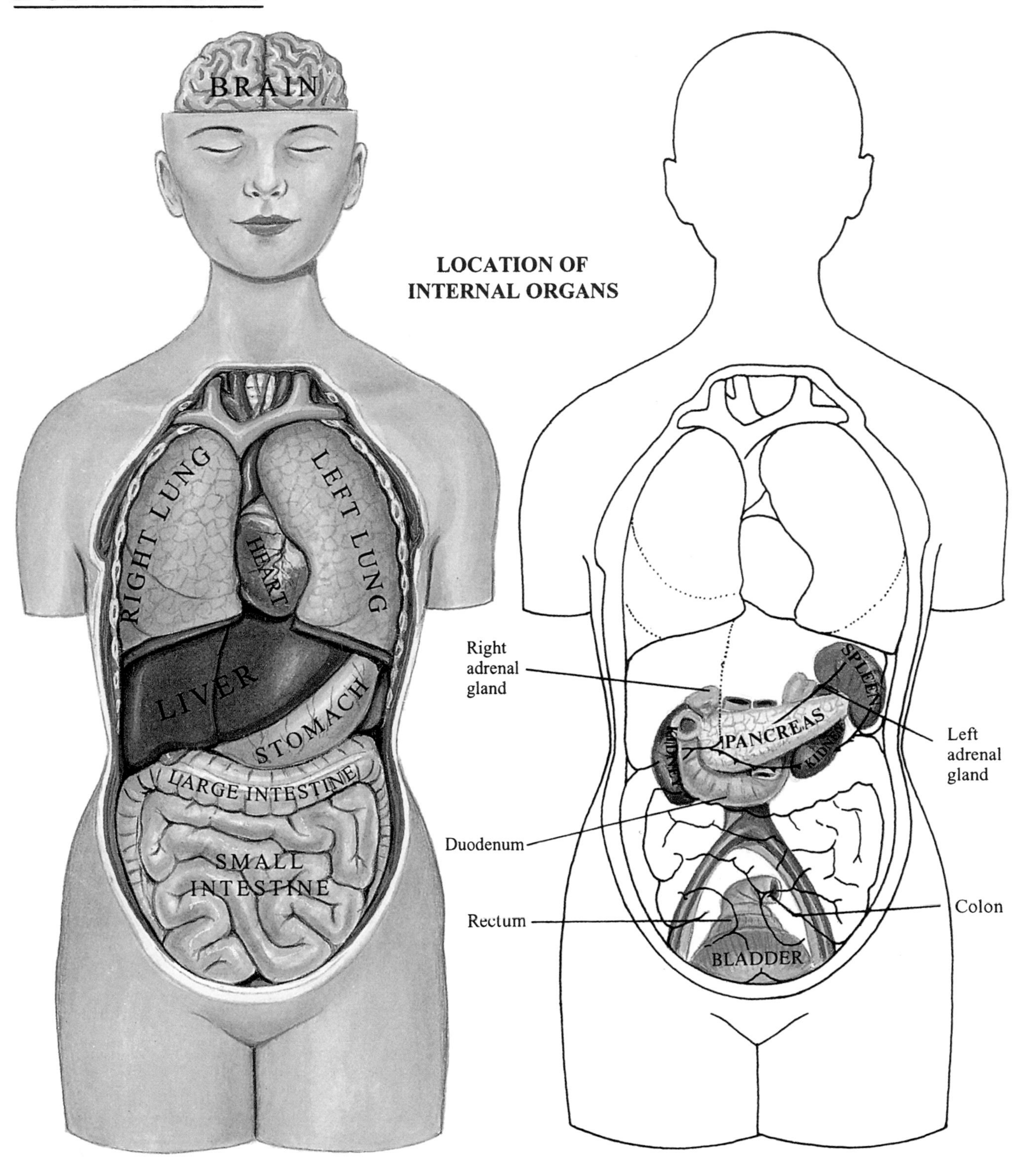

The organs of the body are essential to life. With the exception of the brain, they are situated in the trunk. The two lungs, separated by the heart, fill the upper part of the trunk (thorax). In the lungs, oxygen diffuses into the blood, which is taken to every cell to support life. The heart is the muscular pump circulating blood to all parts of the body.

A sheet of muscle (diaphragm), used in breathing, separates the thorax from the abdominal cavity, which contains the liver, stomach, kidneys, adrenal glands, pancreas, spleen, bladder and intestines. To fit into this space, these organs lie in layers—the intestines, stomach and bladder in front with the kidneys, the adrenal glands, pancreas and spleen behind. The liver lies on top under the diaphragm. Digestion begins in the mouth, where food is chewed and saliva begins to break it down. The process continues in the stomach, the duodenum, the 22-to-25-foot (670-to-760-centimeter) length of coiled small intestine and the inverted U-shaped large intestine, divided into ascending, transverse, descending and sigmoid parts. Finally, food residue passes into the rectum.

The brain is the communications system of the body with overall control over these organs, including glands such as the adrenals and pancreas, which secrete chemical messengers (hormones) to carry out some of its instructions. While the loss of any part of the body is a disadvantage, certain organs are vital. Life is impossible without a functioning brain, heart and liver, and at least one functioning lung and one kidney.

THE HEART

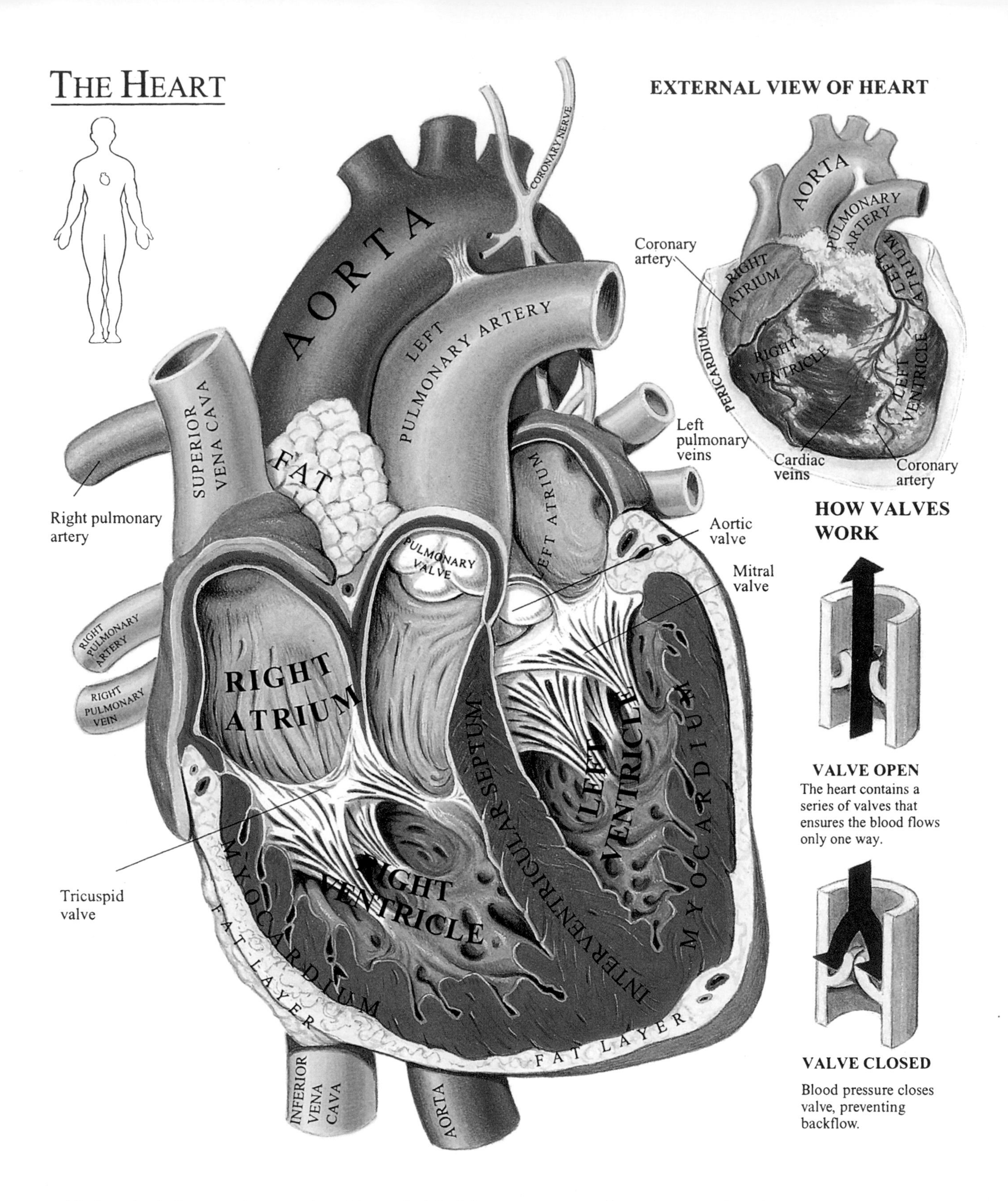

VALVE OPEN
The heart contains a series of valves that ensures the blood flows only one way.

VALVE CLOSED
Blood pressure closes valve, preventing backflow.

The heart is a hollow muscular organ situated just to the left of the breastbone and enclosed in a sac—the pericardium. The heart functions as a pump, circulating the blood through the body. It is separated into right and left halves by the interventricular septum. Each half has a backward-facing receiving chamber (atrium) and forward-facing ejection chamber (ventricle). An orifice controlled by a valve allows blood to pass one way from atrium to ventricle. Deoxygenated blood, high in carbon dioxide, is collected from the body through the two venae cavae (superior vena cava and inferior vena cava) into the right atrium and through the tricuspid valve to the right ventricle, where it is pumped via the pulmonic valve and right and left pulmonary arteries to the lungs, where the blood receives fresh oxygen. The blood then reaches the left atrium through the pulmonary veins. Contraction of the atrium sends the blood through the mitral valve to the left ventricle, whose powerful contractions force blood past the aortic valve into the arch of the aorta and to the rest of the body. The left ventricle has a thicker muscular wall than the right and each contraction corresponds to a pulse felt at the wrist or neck. The heart has its own blood supply—the coronary arteries, arising from the ascending aorta, supply the heart muscle. In a heart attack, these become blocked by a blood clot, with resultant tissue death.

BLOOD CIRCULATION

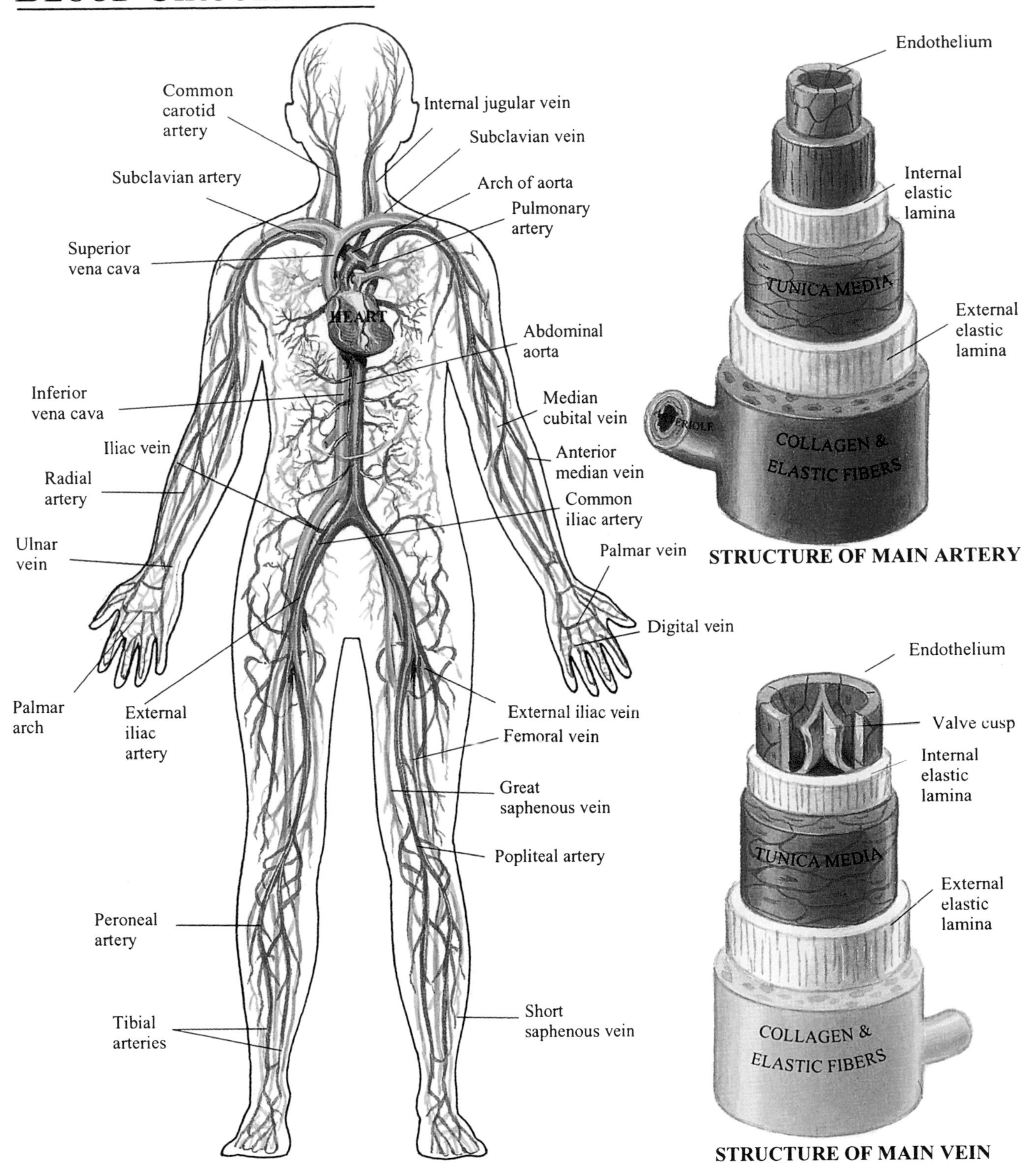

The circulation of the blood provides the transportation system of the body, taking essential materials to every part and removing waste products. Blood is pumped away from the heart by arteries and returned by veins. Except for blood going to the lungs to receive oxygen and lose carbon dioxide (the pulmonary circulation), arterial blood is rich in oxygen and low in carbon dioxide; the opposite is true of venous blood.

Arterial blood leaves the heart via the arch of the aorta, which curves down into the abdomen as the ascending, descending and abdominal aorta. Arteries arise from each part to supply the head, arms and trunk before it divides to supply both legs. Plain muscle in the walls of arteries enables the diameter to vary according to the amount of blood required by the area served. Arteries divide repeatedly into the smallest branches (arterioles), which end in a network of capillaries, with one-cell-thick walls, in organs and tissues. Here oxygen and nutrients are supplied and energy-breakdown products, including carbon dioxide and other wastes, are removed.

Capillaries drain into venules, which combine to form veins ending in both venae cavae. Veins are thinner than arteries, having fewer muscle and elastic fibers in their walls as the pressure is lower in this part of the circulation. Muscular contraction forces blood toward the heart and semilunar cusps in the endothelial lining of veins form "valves" to prevent backflow in the limbs, especially the legs.

Associated with the circulatory system is the lymphatic system, which maintains fluid balance in the tissues and plays a major role in removing bacteria.

The Respiratory System

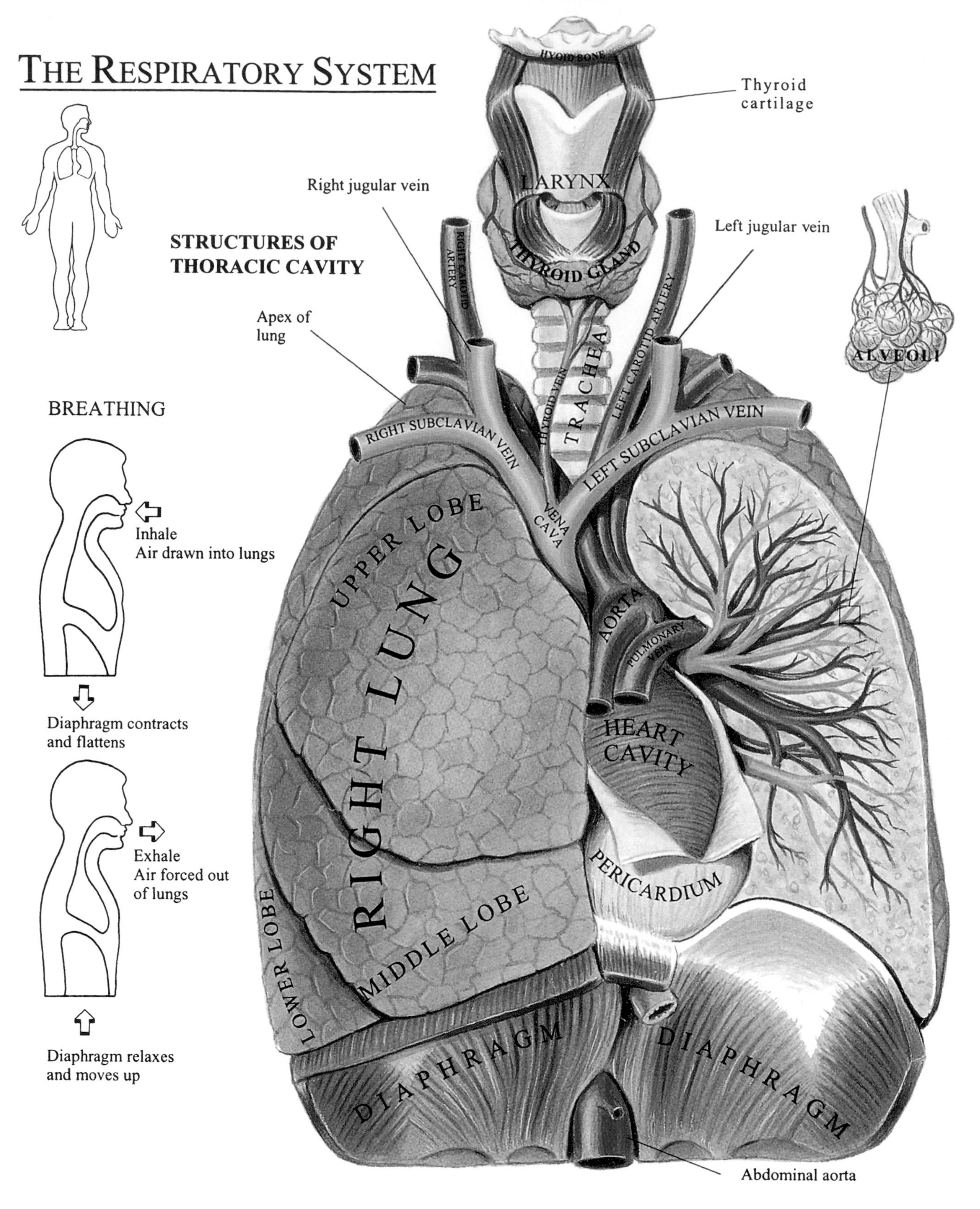

The respiratory system extracts oxygen, constantly required by every living cell in the body, from the air. At the same time carbon dioxide and water vapor, waste products of the energy cycle, are removed from the blood. In breathing, air enters the nose or mouth and is drawn into the larynx (voice box), passing into the chest via the trachea, which divides into a right and left bronchus supplying the two lungs. The right lung, divided into three lobes, is slightly bigger than the left, which has only two. The bronchi repeatedly divide and end in thin-walled sacs (alveoli), surrounded by a network of small blood vessels (capillaries). Blood that is high in carbon dioxide and low in oxygen enters the capillaries from the pulmonary artery and leaves rich in oxygen and low in carbon dioxide, returning to the heart through the pulmonary vein. The lungs have elastic tissue that expands the air sacs as the diaphragm and the chest muscles (intercostals) contract, increasing the internal chest area and drawing air into the lungs through the trachea. Exhalation is the reverse of this process. Breathing is under the control of the respiratory center in the brain, which responds to concentrations of carbon dioxide in the blood. Holding one's breath causes the level of carbon dioxide in the blood to rise and the breathing center overrides willpower, forcing one to gasp for breath. The mucous lining of the respiratory tract produces sputum. Irritants affect the lining and cause a "cough" to expel the foreign material.

Digestive System

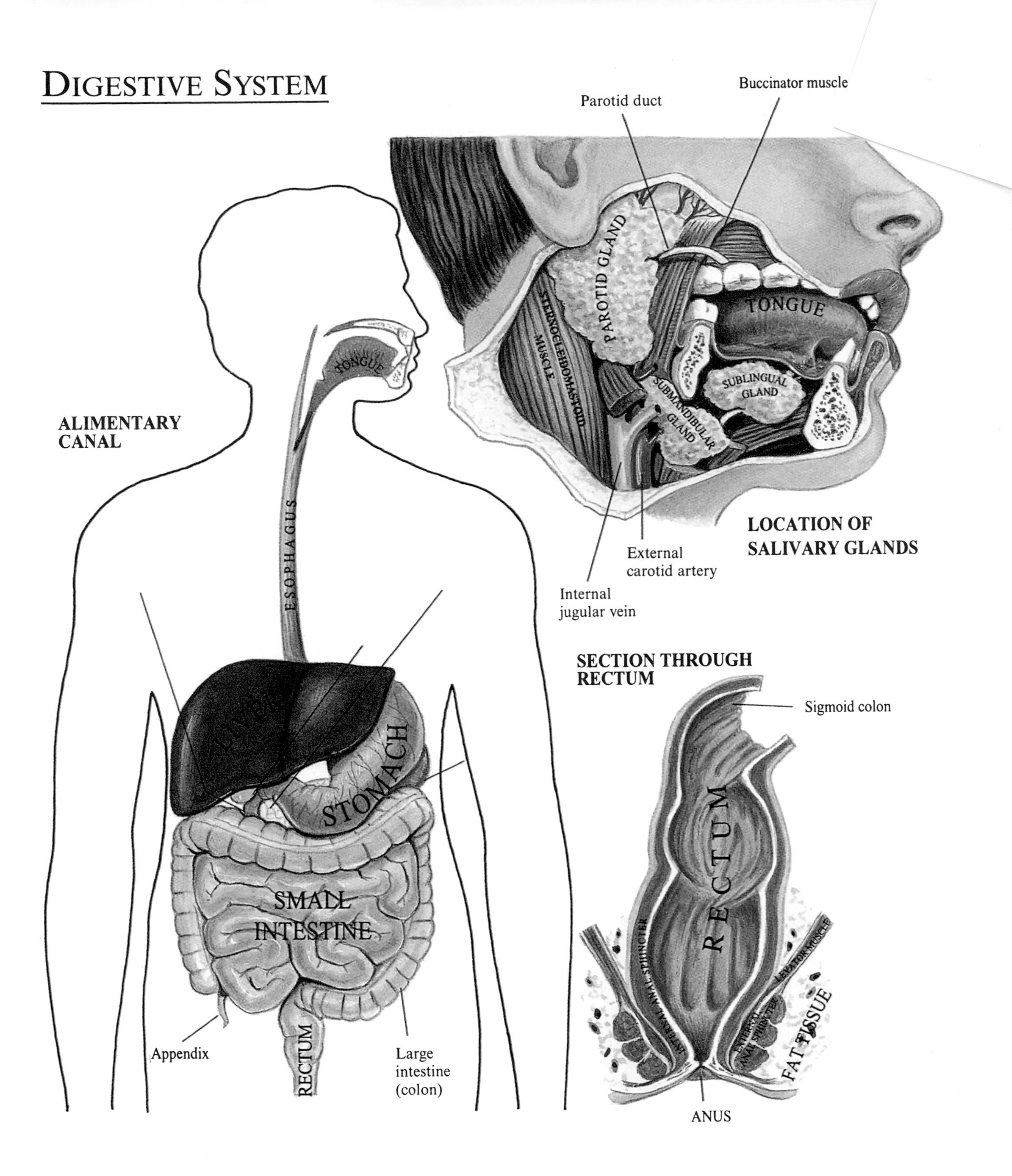

The digestive system breaks down food into a form that can be absorbed by the blood and transported to all parts of the body to provide energy and nutrients. Muscular waves in the gut wall (peristalsis) propel food through the various processing tubes. Strong contractions are felt as abdominal pain ("colic") or as a message to empty the rectum by relaxing the muscular valve (anal sphincter) and passing waste products of digestion (feces) out of the body.

Digestion begins in the mouth, where food is moved around by the tongue and chewed by the teeth, using the buccinator muscles. Three pairs of salivary glands (sublingual, submandibular and parotid) supply saliva, which contains an enzyme that lubricates and begins digestion of starch in food. Other enzymes farther down the gut have specific actions on carbohydrates, fats or proteins.

The chewed mixture passes as a mass (bolus) down the food pipe (esophagus) from the mouth to the stomach, where it is churned into a gruel-like substance called chyme by the action of hydrochloric acid, water, mucus and pepsinogen. The acid turns the pepsinogen to pepsin, which digests protein. The chyme stays in the stomach from 30 minutes to three hours. Starchy snacks stay for a shorter time than fatty meals. As acid levels rise, the muscular valve (pyloric sphincter) from the stomach relaxes, slowly releasing chyme into the small intestine.

STOMACH & INTESTINE

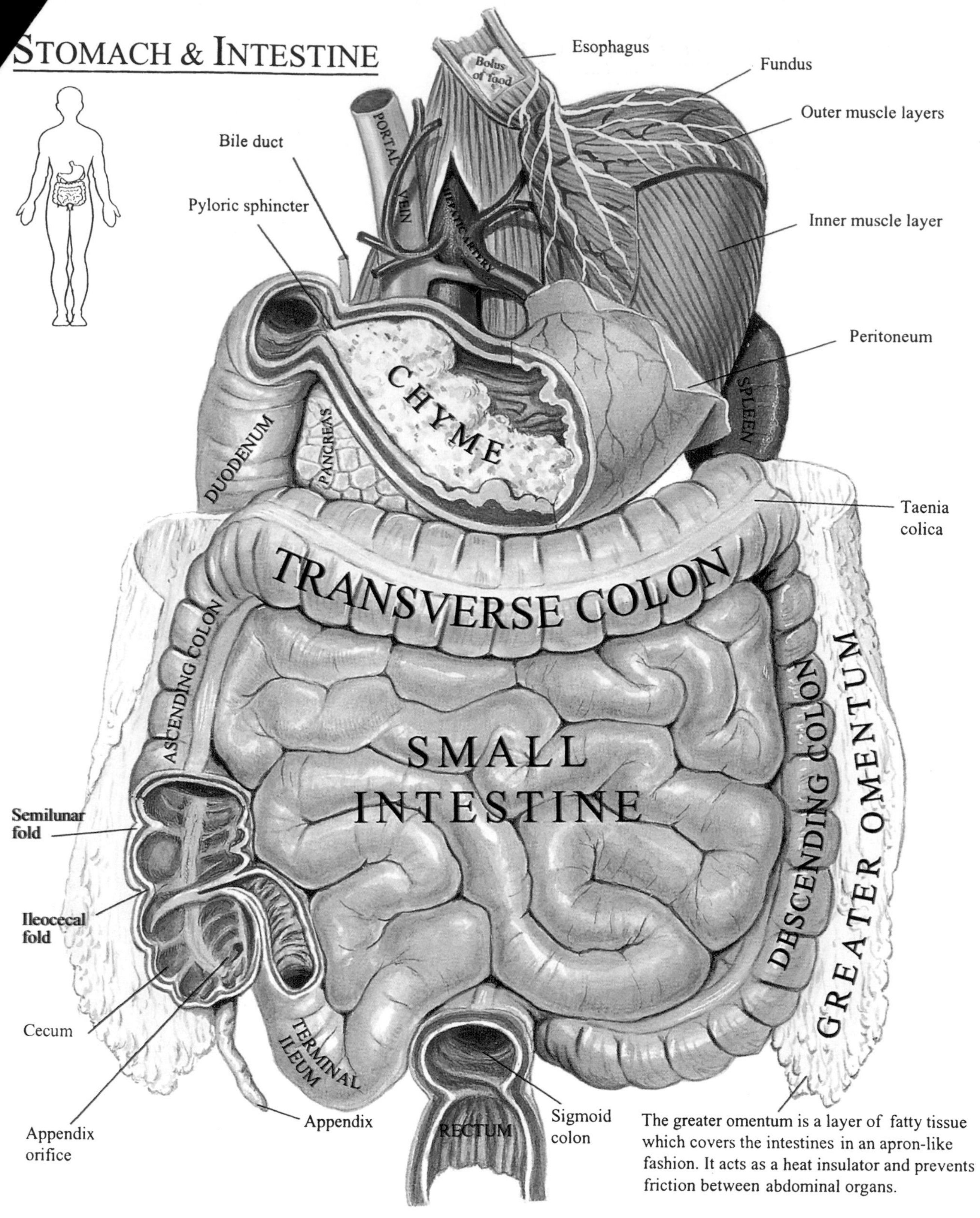

The greater omentum is a layer of fatty tissue which covers the intestines in an apron-like fashion. It acts as a heat insulator and prevents friction between abdominal organs.

The small intestine, 19½ feet (six meters) long, fills the central abdomen below the liver and stomach. The first 9¾ inches (25 centimeters) is called the duodenum. Here the chyme is mixed with bile and pancreatic juices. Alkaline bile salts emulsify fat, increasing the surface area for enzyme action. Bile is stored in the gallbladder. Pancreatic secretions contain four enzymes: fat-splitting lipase; starch-breaking amylase and two that continue protein breakdown, but are activated only when they meet enterokinase, one of the six enzymes in the intestinal juice. Thus, carbohydrates are broken down into simple sugars such as glucose; proteins, into amino acids; and fats, into fatty acids and glycerol. All these are removed by the blood or lymphatic system. The intestinal wall provides a huge surface area for absorption, being folded and provided with villi, finger-like projections containing blood vessels and lacteals—lymphatic vessels originating in the villi.

The last part of the small intestine, the ileum, joins the large intestine at the cecum, ending in a narrow, blind tube, the appendix, ¾ to 7¾ inches (2–20 centimeters) in length. Performing no known function in man, it is thought to be an evolutionary remnant. It sometimes becomes inflamed (appendicitis). The large intestine, 59 inches (1.5 meters) long, forms an arch, with ascending, transverse and descending portions, ending as the sigmoid colon and rectum in the pelvis. Salts and water are absorbed in the large intestine, so that the feces, comprised of indigestible products, are formed and not fluid. The intestines and abdominal cavity are covered by a moist transparent sheet (peritoneum) that enables coils of intestine to slide over one another.

THE LIVER, PANCREAS & SPLEEN

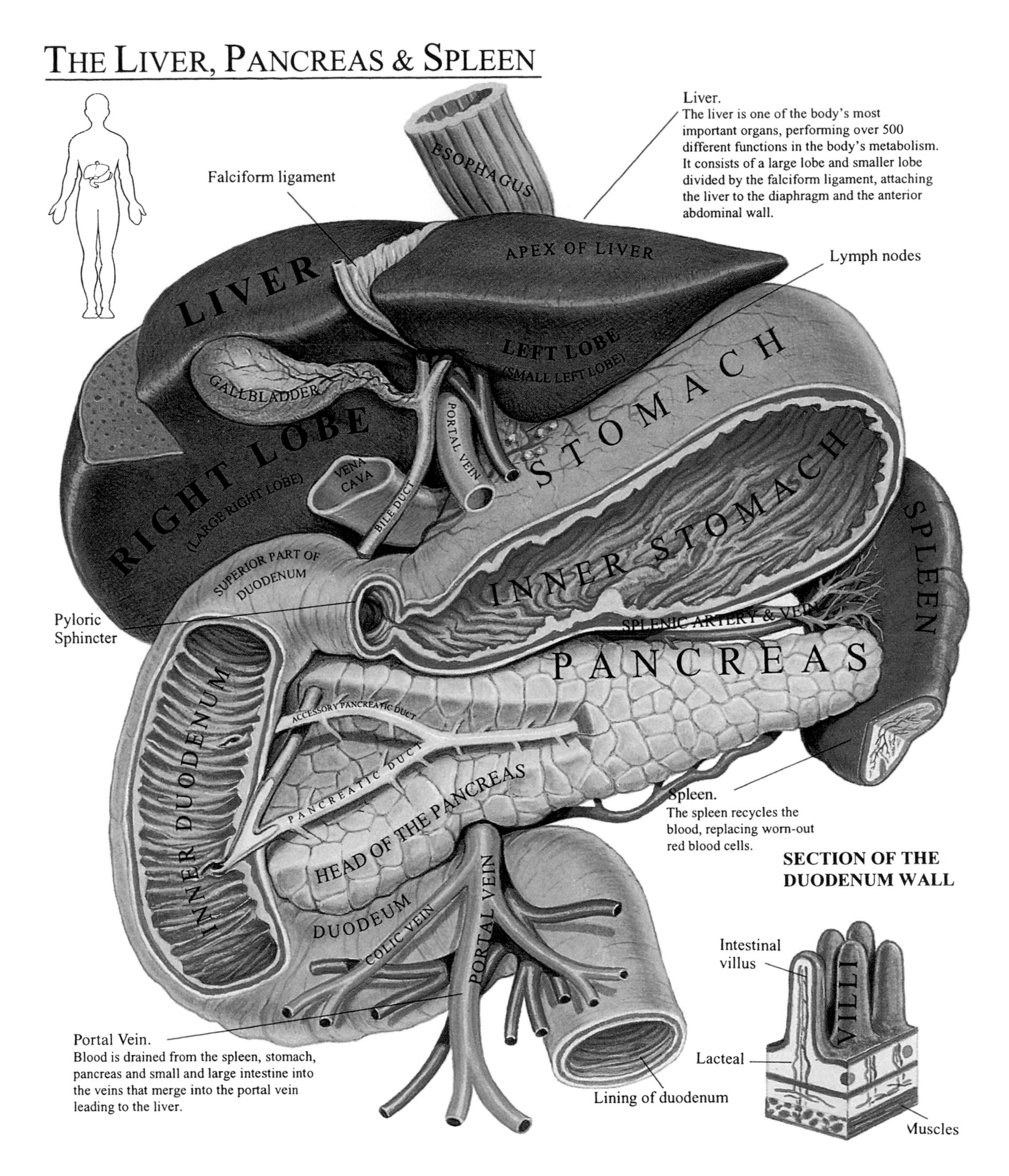

The liver, pancreas and spleen are three major organs nestled under the diaphragm. The liver is the largest internal organ, weighing about 3 pounds (1.4 kilograms). Blood, rich in nutrients collected from the intestinal villi, reaches the liver via the portal vein and is processed to break down fats, use amino acids for tissue growth and repair, convert protein waste into urea that is excreted by the kidneys, detoxify drugs and poisons and produce important blood-clotting factors. Vitamins A, B12 and D, iron and excess glucose converted by the liver into glycogen are all stored in the liver. Each day about one liter of bile is produced by liver cells, collected and stored in the pear-shaped gall-bladder under the right lobe of the liver. The yellow-green color of bile is due to the breakdown products of hemoglobin in red blood cells. Bile, containing salts that emulsify fats in food, reaches the upper part of the duodenum via the bile duct. The pancreas, about 6 inches (15 centimeters) long, lies under the left lobe of the liver and produces pancreatic enzymes that aid digestion. Islands of cells in the pancreas secrete a protein hormone—insulin—that controls glucose metabolism and blood-sugar levels. Failure of insulin production by the pancreas causes diabetes. The spleen, composed of lymphoid tissue and red cells, weighs 8 ounces (220 grams). The spleen produces white blood cells and antibodies that combat infection, destroys worn-out red blood cells and filters out bacteria.

THE URINARY SYSTEM

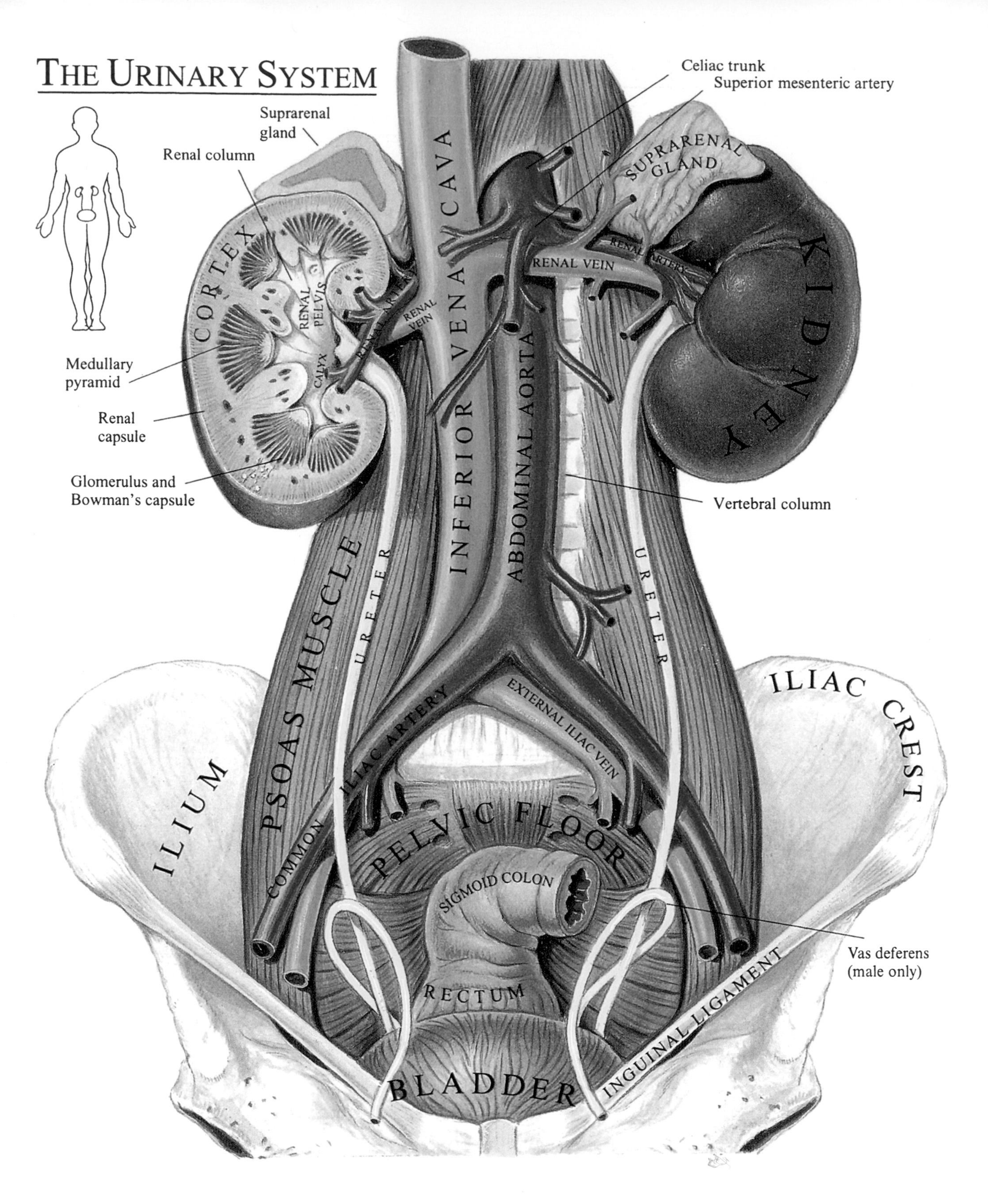

The urinary system consists of two kidneys, each with a ureter conveying urine from the kidneys to a collecting sac called the bladder, and a pipe, the urethra, leading to the exterior. The bean-shaped kidneys, 4¼ inches (11 centimeters) long and 1⅛ inches (3 centimeters) thick, lie on either side of the spine, level with the upper lumbar vertebrae. Inside lies a pale outer cortex and a darker medulla composed of tubules leading into a collecting space called the renal pelvis. Kidney tissue is composed of over one million coiled tubules called nephrons. Each begins in the cortex as a cup-shaped Bowman's capsule, containing a tuft of capillaries known as a glomerulus. Convolutions increase the length of a nephron, and a large loop bends back into the cortex to meet more capillaries from the renal artery, before ending in the medulla. Thus, blood passes through two sets of capillaries in one organ, something that occurs nowhere else in the body. Waste products are filtered out at high pressure—about 162 fluidrams (600 milliliters) of blood per minute pass through the kidney—to maintain the balance of chemicals in the blood. To fine-tune this process, chemicals are selectively reabsorbed or secreted in the end tubules to produce urine, composed of 96 percent water and 4 percent salts and waste products.

Male Reproductive System

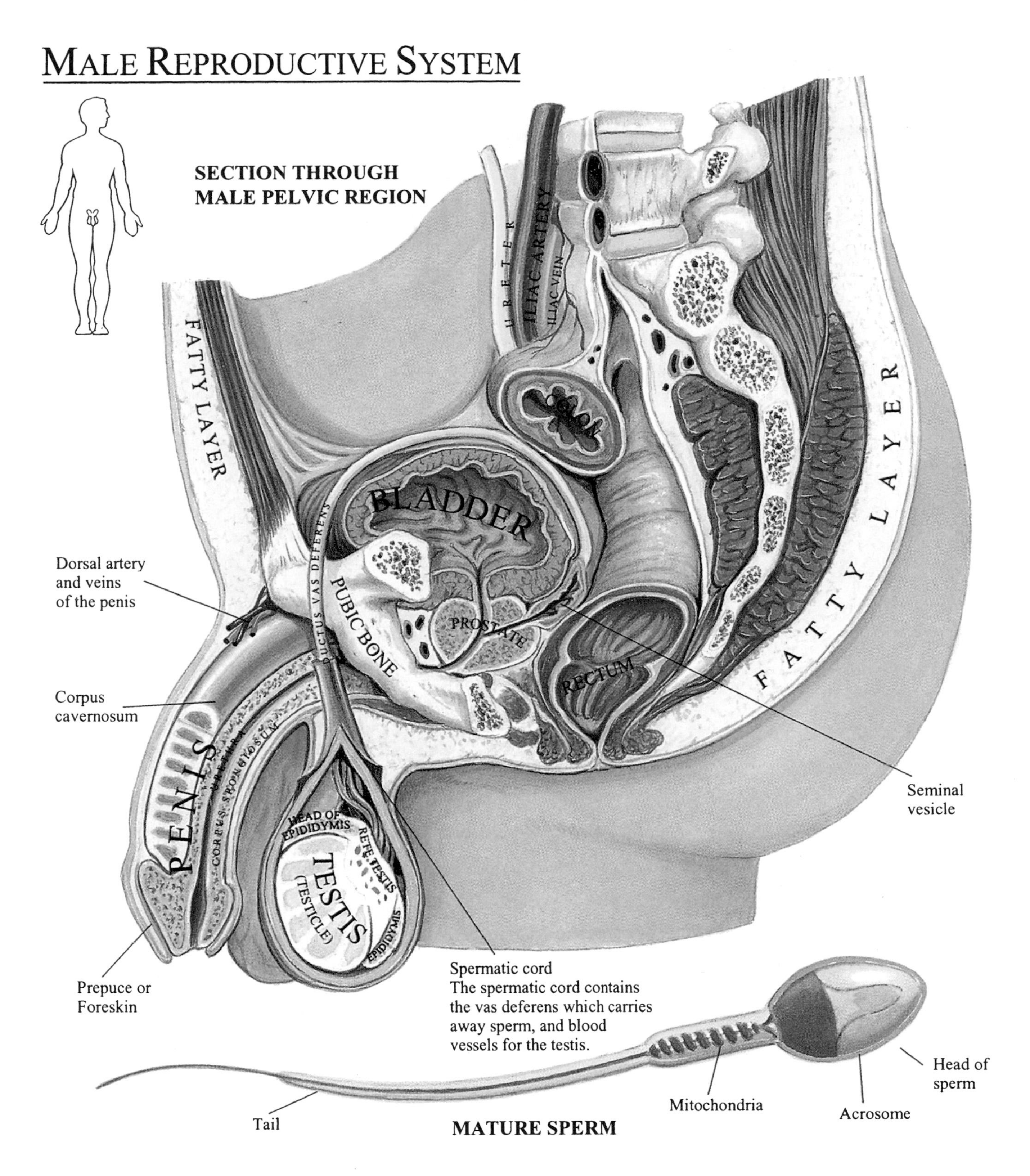

The male reproductive system visible on the exterior of the body consists of two testes suspended in a sac (scrotum) and a penis. The main reproductive glands (testes) consist of 200–300 lobules containing 1–3 tiny, coiled tubules, the walls of which contain cells that develop into spermatozoa; the connective tissue contains cells secreting the male hormone, testosterone. After puberty, up to 200 million sperm are produced daily. It is the sperm, contained in a liquid called semen, that fertilizes the ovum (egg) in the female. Each sperm has a head containing a nucleus of genetic material, a middle, and a tail enabling it to swim toward an ovum. (The sperm shown here is greatly magnified. The average length of the head is about 5 microns wide; the tail is about 50 microns long. A micron is about 1/25,000th of an inch.) The seminiferous tubules open into the tightly coiled epididymis, which runs down the back of the testis to the vas deferens—a duct running between the base of the bladder and rectum to the prostate gland, which surrounds the urethra below the bladder. There the vas deferens joins the duct of the seminal vesicles—two pouches secreting an alkaline fluid that feeds sperm and constitutes most of the seminal fluid. These tubes fuse to form the ejaculatory ducts, which pass through the prostate gland and open into the urethra. The urethra opens to the exterior at the tip of the penis—a tubular organ containing tissue capable of swelling with blood to elongate and harden (an erection), allowing insertion into the vagina. The urethra has a dual purpose—to excrete urine and to carry sperm when ejaculation occurs. The entrance to the bladder is closed during sexual activity to prevent urine from killing sperm.

Female Reproductive System

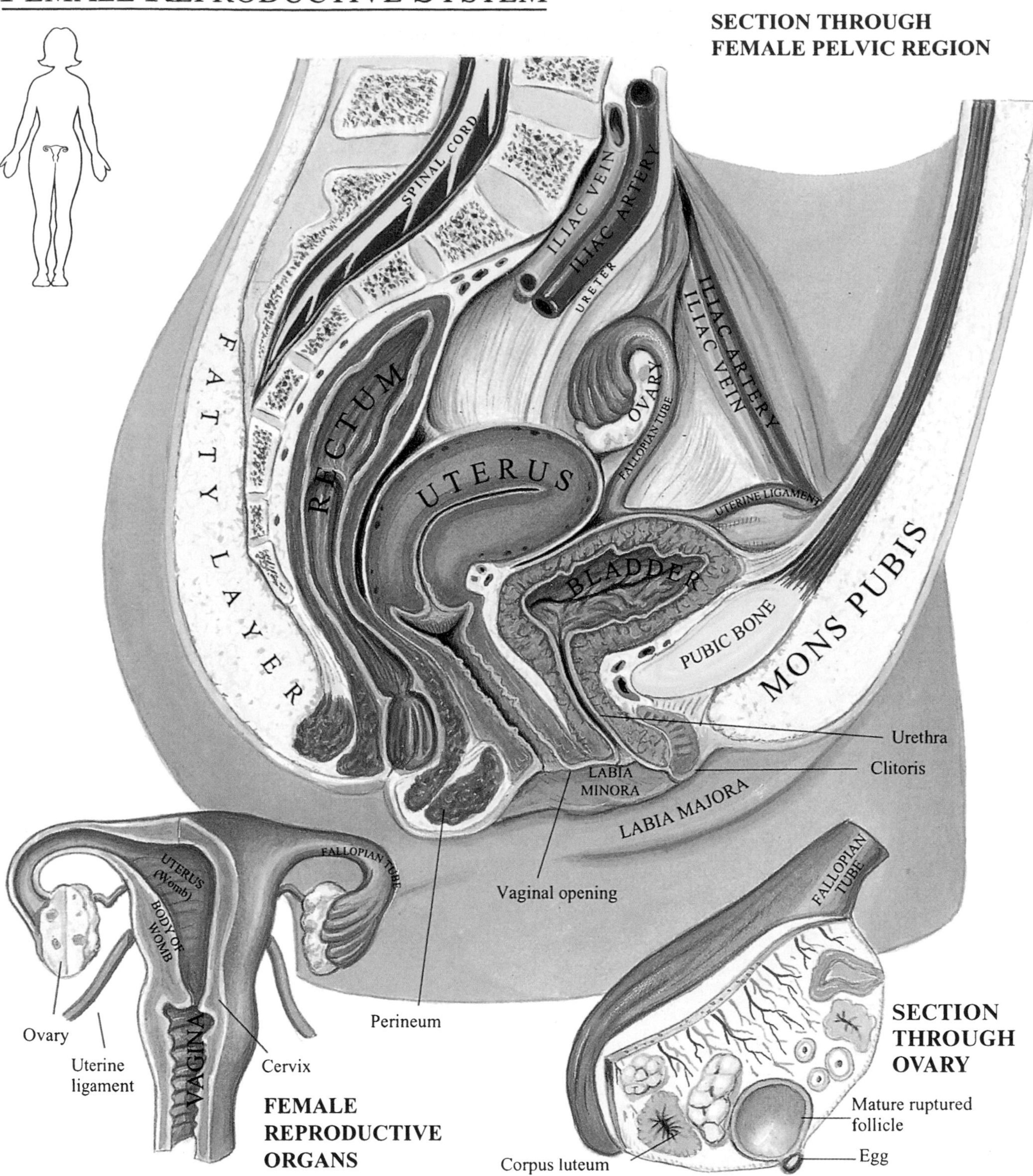

The female reproductive system consists of internal and external organs. The vaginal opening and associated glands, together with the surrounding fleshy lips (labia majora and minora), clitoris (a small, sensitive mound of erectile tissue corresponding to the male's penis) and the fatty pad of the mons pubis form the external genital organs. The urethra opens between the clitoris and vagina. The perineum is the area of skin between the vagina and anus overlying the muscles of the pelvic floor.

The internal female genital organs are the ovaries, fallopian tubes, uterus (womb) and the opening to the exterior, known as the vagina. The ovaries are two small glands, the size of almonds, suspended on either side of the uterus by ligaments. The ovaries start to function at puberty. The ovary produces eggs (ova), one of which matures each month in the vesicular or graafian follicle, which ruptures, releasing an egg. The egg passes through the frilled opening into the uterine tube, where it may be fertilized by a male sperm after intercourse. The empty follicle changes into specialized tissue (corpus luteum) which remains active by producing a hormone, progesterone, to prepare the womb for the fertilized egg if a pregnancy occurs. The uterus is a hollow, thick-walled, muscular organ in which a baby develops. It has a vascular lining rich in blood. If it does not receive a fertilized egg, this lining is shed each month (menstruation).

Development of a Baby

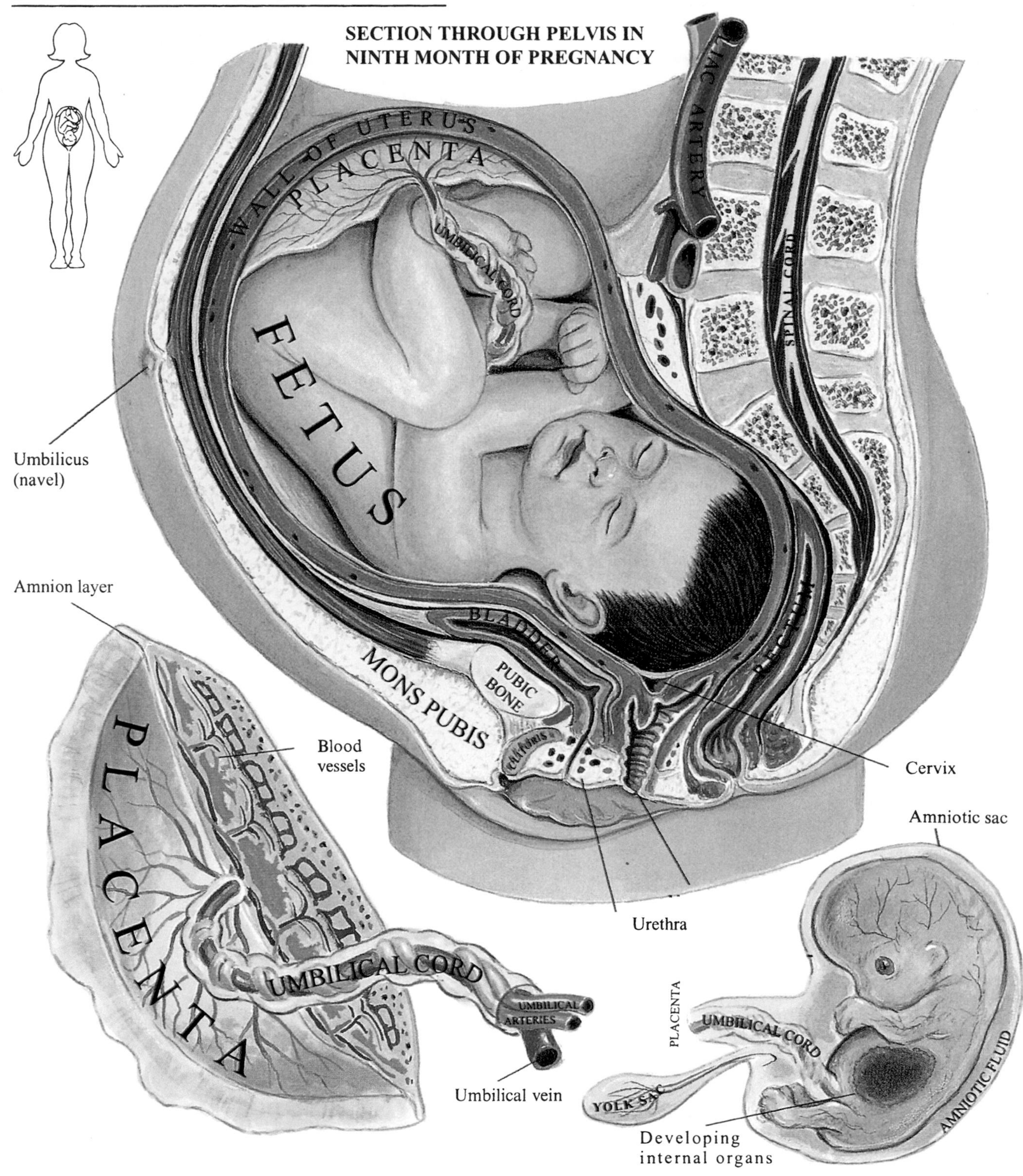

A baby, weighing an average of 7½ pounds (2.8 kilograms) at birth, begins life nine months earlier as a single cell, formed when a female egg is fertilized by a male sperm in the fallopian tube. The sperm and the egg each contribute half the inherited characteristics of the new baby. The sperm carries the chromosome that determines its sex.

The fertilized egg repeatedly divides and implants in the lining of the uterus (womb). The new cells begin to differentiate into tissues and organs. A fetus, with a recognizable head, eyes, heart and limbs, enclosed in a fluid-filled sac (amniotic sac), is joined by the umbilical cord and a mesh of blood vessels (placenta) to the part of the uterine wall where embedding took place. The placental blood vessels are in close contact with uterine maternal vessels so that the fetus receives oxygen, food and protective antibodies and can give up waste products.

Twelve weeks after conception, the fetus, only about 2¹⁄₁₆ inches (56 millimeters) long, is fully formed. As the baby grows, the uterus enlarges to accommodate it. Between 16 and 22 weeks, the mother becomes aware of the baby's movements. The enlarged womb begins to show as an abdominal swelling. At full term (40 weeks) the mature baby, usually lying head down, is ready for birth; the amniotic sac ruptures and strong uterine contractions push the baby through the open cervix and vagina into the outside world, followed by the placenta. The umbilicus (navel) marks the site of the severed umbilical cord.

INDEX

References to illustrations are in *italics.*

HUMAN ANATOMY in full color

JOHN GREEN and JOHN W. HARCUP

The human body is an infinitely complex marvel of fine design, superbly adapted to its functions. A host of specialized organs, bones, muscles, nerve fibers, blood vessels and other anatomical features work together in harmony to maintain the network of interrelated body systems necessary to maintain life. Now the component parts of this intricate flesh-and-blood machine are clearly revealed in this treasury of detailed anatomical illustrations.

Noted illustrator John Green has rendered 25 exceptionally clear and precise full-color plates of the body's organs and systems: the skeleton, muscles and skin, as well as the respiratory, digestive, circulatory, reproductive and other systems. Illustrations also focus on such important organs as the eye, ear and brain. Each carefully labeled plate has been reviewed for accuracy and is accompanied by an extensive caption written by Dr. John W. Harcup, clearly explaining the nature and purpose of the body part or system represented.

Its precision and clarity make this book an ideal supplement to school courses in biology, health and other subjects, but it will also appeal to general readers, who will enjoy its wealth of superb illustrations illuminating the incredibly complex and highly specialized workings of the human body.

www.doverpublications.com